Recent Advances in Immunoassays

I0487504

Samar K Kundu
August 5, 2014

Recent Advances in Immunoassays

Table of Contents

Recent Advances in Immunoassays

1. Introduction

Immunoassay is a biochemical method that measures the presence or concentration of an analyte referred as an *antigen* in a solution through the use of an antibody or immunoglobulin. The assay relies on the ability of an antibody to recognize and bind a specific area of the antigen called an *epitope* in a complex mixture of molecules. Similarly constructed assays may use specific binding proteins, such as transport proteins or receptors that are not antibodies. The term *ligand-binding assay* is used in such cases.

In an immunoassay, the analyte is bound to the antibody forming an immune complex. This complex is separated from the unbound reagent fraction by physical or chemical separation technique. Analysis is achieved by measuring the label activity (e.g. radiation, fluorescence, chemiluminescence or enzyme) in either of the bound or free fraction. A standard curve, referred as a *calibration curve* which represents the measured signal as a function of the concentration

Recent Advances in Immunoassays

of the analyte in the sample is constructed. Unknown concentration of the analyte is determined from this calibration curve.

Immunoassay methods have been widely used in many areas of in the clinical diagnosis, screening, disease monitoring, bacterial, viral, therapeutic drug monitoring, biopharmaceuticals, pesticides and many other areas (1-4). The analysis in these areas usually involves measurement of very low concentrations of macromolecules, biomolecules, low molecular weight drugs, metabolites, and/or biomarkers which indicate disease diagnosis or prognosis. The importance and widespread of immunoassay methods in pharmaceutical analysis are attributed to their inherent specificity, high-throughput, and high sensitivity for the analysis of wide range of analytes in biological samples. The detection system in immunoassays depends on readily detectable labels (e.g. radioisotopes or enzymes) coupled to one of the immunological reagents (i.e. analyte or antibody). The use of these labels in immunoassays results

Recent Advances in Immunoassays

in assay methods with extremely high sensitivity and low limits of detection (LOD). In circumstances, where the specific measurements of large molecules at the femtomole to attomole level in complex biological matrices are required, immunoassays are the methods of choice because of their high specificity and sensitivity.

In any immunoassay method, the antibodies are the key reagents on which the success of any immunoassay depends. The antibodies can be either polyclonal or monoclonal. However, for immunoassay development for pharmaceutical analysis purposes, monoclonal antibodies are more advantageous than polyclonal ones. This is attributed to their higher degree of affinity and specificity towards the analyte. Even that, many successful immunoassays were developed using polyclonal antibodies because it is possible to generate the antibodies with high affinity to the analyte (5).

Recent Advances in Immunoassays

The signal generating labels in immunoassays include radioactive atoms (mostly 125I, 3H, and 14C) (6, 7). The use of radioactive labels offers extremely sensitive and quite precise assays; however, they have some draw backs (e.g. health hazards, special attention for handling of the reagents, training of staff, short half-life time of the isotope) and expensive instrumentation for the counting of radioactivity. Therefore, alternative non-radioactive labels such as enzymes (8-10), fluorescent probes (11, 12), chemiluminescent substances (13-21), metals and metal chelate (22, 23), and liposomes (24, 25) were introduced. On the basis of number of publications, enzymes are the most common labels employed in immunoassay methods. A potential advantage in the use of enzyme labels for immunoassay is their ability for amplification of the signal, and subsequently the potential for increasing in the sensitivity of the assay. This is beneficial and allows attending the desirable sensitivity for the analysis

Recent Advances in Immunoassays

The range of design possibilities for immunoassay is multidimensional, but in general can be classified into heterogeneous or homogeneous assay, respectively. These methods can be performed in either competitive or non-competitive designs. The choice from these designs is based on nature of the antigen, antibody, labeling chemistry available and the analytical parameter required from the assay (e.g. sensitivity, specificity, dynamic range, precision).

2. Enzyme Linked Immunosorbent Assays

Since the 1970s, the enzyme linked immunosorbent assay (ELISA) has been the standard against which immunoassay performance is measured. ELISAs have been developed in many formats (26).In the most common type of non-competitive immunoassays, the antigen of interest is immobilized by direct adsorption or attaching a capture antibody on to a solid surface. Detection of the antigen can then be

Recent Advances in Immunoassays

performed using an enzyme-conjugated primary antibody (direct detection) or a matched set of unlabeled primary and conjugated secondary antibodies (indirect detection). This is shown in Figure 1,

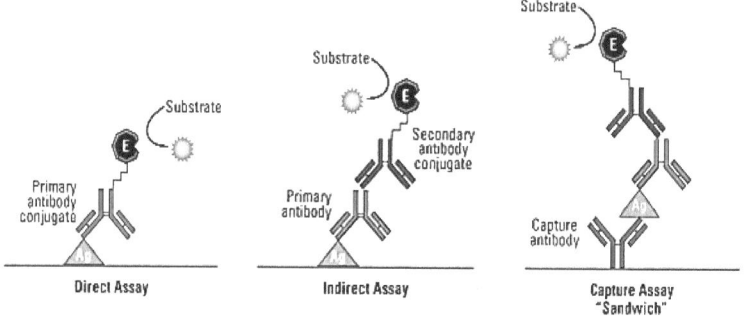

Figure 1. Common Non-Competitive Immunoassays

The direct detection method uses a labeled primary antibody that reacts directly with the antigen. Direct detection can be performed with antigen that is directly immobilized on the assay plate or with the capture assay format. Direct detection is not widely used in ELISA but is quite common for immunohistochemical staining of tissues and cells. The indirect detection method uses a labeled secondary antibody for detection

Recent Advances in Immunoassays

and is the most popular format for ELISA. The secondary antibody should have the specificity for the primary antibody. In a sandwich ELISA, it is critical that the secondary antibody be specific for the detection primary antibody only (and not the capture antibody) or the assay will not be specific for the antigen. Generally, this is achieved by using capture and primary antibodies from different host species (e.g., mouse IgG and rabbit IgG, respectively). For sandwich assays, it is beneficial to use secondary antibodies that have been cross-adsorbed to remove any antibodies that have affinity for the capture antibody.

In the assay, the antigen of interest is immobilized by direct adsorption or attaching a capture antibody on to a solid surface. Detection of the antigen can then be performed using an enzyme-conjugated primary antibody (direct detection) or a matched set of unlabeled primary and conjugated secondary antibodies (indirect detection).

Recent Advances in Immunoassays

An immunoassay can also be performed as a competitive assay. This is common when the antigen is small and has only one epitope, or antibody binding site. One variation of this method consists of labeling purified antigen instead of the antibody. Unlabeled antigen from samples and the labeled antigen compete for binding to the capture antibody. A decrease in signal from purified antigen indicates the presence of the antigen in samples when compared to assay wells with labeled antigen alone.

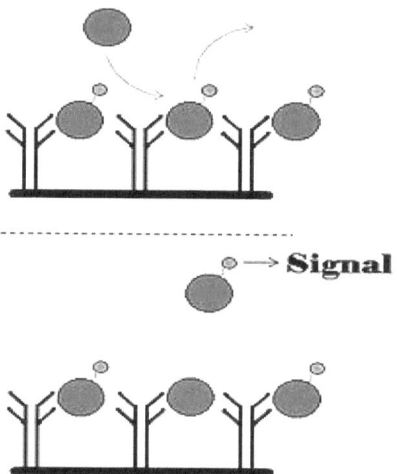

Figure 2. Competitive Heterogeneous and Homogeneous Immunoassays

Recent Advances in Immunoassays

Immunoassay methods that require separation of bound Ab-Ag* complex are referred to as heterogeneous immunoassays. For separation of Ab-Ag* complex, solid phase reagent such as magnetic microparticles or plastic beads are utilized. Those that do not require separation are referred to as homogeneous immunoassays. On the other hand, homogeneous methods have been generally applied to the measurement of small analytes such as abused and therapeutic drugs. Since homogeneous methods do not require the separation of the bound Ab-Ag* from the free Ag*, they are generally much easier and faster to perform.

ELISAs that detect biological agents are heterogeneous assays (shown in Figure 2) in which an agent or an agent-specific antigen is captured onto a plastic microtiter plate or microparticles by a "capture" antibody previously bound to the solid matrix. Bound antigen is then detected using a secondary "detector" antibody. The detector antibody can be directly labeled with

Recent Advances in Immunoassays

a signal-generating molecule, or it can be detected with another antibody that is labeled with an enzyme. These enzymes catalyze a chemical reaction with a substrate that results in a colorimetric change. The intensity of this color can be measured by a spectrophotometer, which determines the optical density of the reaction, using a specific wavelength of light. Many ELISA formats require antibodies from two different species of animals so there is no direct interaction between sandwich layers. If the detector antibody were directly labeled with enzyme, antibodies from the same species could be used as both capture and detector reagents

3. Signal Detection

Assay sensitivity is also dependent on the signal used for measurement. Depending on the nature of the signal, the reactants may be detected visually, electronically, chemically, or physically, and a wide range of instruments can detect the presence of these labels with a high degree of sensitivity. Enzymes are effective labels because

Recent Advances in Immunoassays

they catalyze chemical reactions, which can produce a signal. Because a single enzyme molecule can catalyze many chemical reactions without being consumed in the reaction, these labels are effective at amplifying assay signals (27). Most enzyme-substrate reactions used for immunoassays utilize chromogenic, chemiluminescent, or fluorescent substrates that produce a signal detectable with the naked eye, a spectrophotometer, luminometer, or fluorometer (27-31). A disadvantage of enzyme-based assays is that both the enzymes and substrates may be unstable and require specialized storage to maintain activity. Fluorescent dyes and other organic and inorganic molecules capable of generating luminescent signals are also commonly used labels in immunoassays (29). Assays using these molecules are often more sensitive than enzyme-based immunoassays but require specialized instrumentation and often suffer from high background contamination due to the intrinsic fluorescent and luminescent qualities of some proteins and light-scattering effects. Signals for

Recent Advances in Immunoassays

assays with these types of labels are amplified by integrating light signals over time and cyclic generation of photons. Other commonly used labels include gold, latex, and magnetic or paramagnetic particles. All are quite stable under a variety of environmental conditions and can be detected by visual inspection or instruments. However, these labels are essentially inert and therefore do not produce an amplified signal. Signal amplification is useful and desirable because it results in increased assay sensitivity. Increased signal strength can be attained by using amplifiable labels as described above or by using molecules capable of forming multiple bonds. These molecules can produce more complex lattices of signal-generating compounds or molecules. Biotin and avidin are examples of molecules exhibiting these characteristics. They have very high affinities for each other, developing almost irreversible bonds (Kd = 10-15 M). In addition, avidin can bind as many as four biotin molecules, increasing the size of the complex. If biotin is bound to a signal-generating molecule or

Recent Advances in Immunoassays

compound, the strength of the signal increases proportionally.

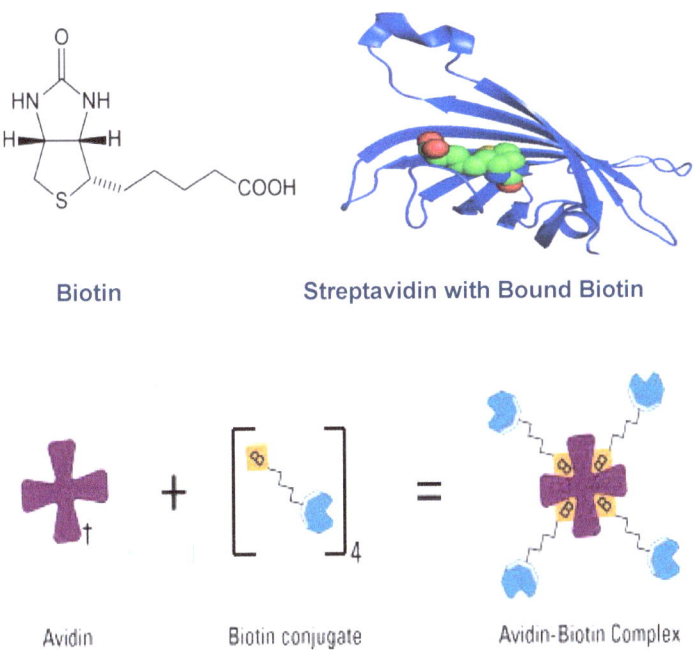

Figure 3. Signal Amplification with Biotin-Streptavidin in Sandwich ELISA

4. Enzyme Detection

Two enzymes are commonly used in ELISA applications. Horseradish Peroxidase (HRPO),

Recent Advances in Immunoassays

found in the roots of plant horseradish, is a more commonly used enzyme. Its small size (40 KDa) allows more molecules to be coupled to antibodies or avidin, and this can boost signal generation.

HRPO is the enzyme of choice for most researchers performing ELISAs and can be used with a variety of substrates, most of which are more sensitive than AP equivalents. In recent years, PolyHRPO is being in immunoassays because it delivers an enhanced enzymatic label comprising covalent horseradish peroxidase polymer. PolyHRPO conjugates deliver a large number of signal-generating enzyme molecules to one bound analyte molecule. This results in multiple detection enhancements which are directly proportional to the HRP polymerization range resulting in easy detection of low-abundance targets (low pictogram-femtogram range) in immunoassays, e.g. Streptavidin-PolyHRPO conjugates are made of 5 (five) identical covalent HRP homopolymer blocks that

Recent Advances in Immunoassays

are, also covalently, coupled to multiple streptavidin molecules (32, 33).

Alkaline Phosphatase (ALP) is a large enzyme used in a minority of assays. Two types of ALP are used in immunoassays. One is calf-intestinal alkaline phosphatase (CIAP/CIP) is a dimer with molecular weight is 140,000 and the other one from *E.coli* with molecular weight of 80,000. They differ quite substantially and should not be substituted by other for an immunoassay. The size of ALP makes it difficult to conjugate more than one or two molecules of the enzyme to each molecule of an antibody or avidin, and this limits the amount of signal that can be generated. ALP is also prone to stability issues unless stored and handled correctly.

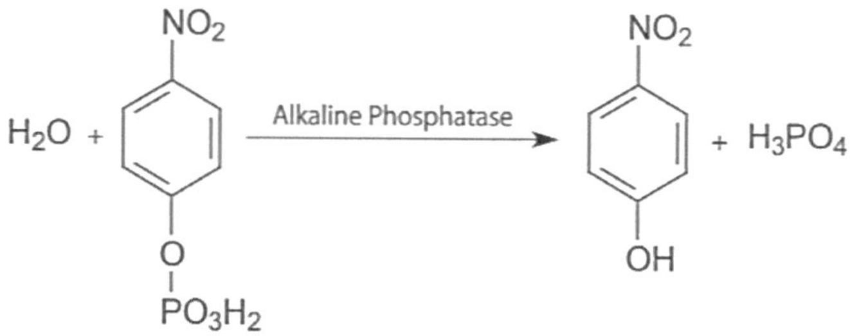

Recent Advances in Immunoassays

Figure 4. Dephosphorylation of a Phosphate Substrate by Alkaline Phosphatase

Another large tetramer enzyme ß-Galactosidase from *E. coli* with molecular weight 464 Kda has been used in immunoassays but its use is limited and it is not being covered in this book.

5. Type of Label (Substrates)

Since the demise of an early generation of automated RIA instruments, dedicated clinical laboratory analyzers have used only non-isotopic labels. Non-isotopic immunoassay most commonly uses some type of optical signal as an end point: absorbance, light scatter, fluorescence, or chemiluminescence. Of these, absorbance is the

Recent Advances in Immunoassays

least sensitive. Since concentrations much below 1 µmol/L cannot be measured, Chromophores cannot be used directly as labels, but an enzyme label can be detected via a chromogenic substrate. Absorbance measurements are subject to positive interference, e.g., with hyperlipidemic, or hemolyzed samples, but this may be mitigated by washing steps or kinetic measurements. Absorbance is not readily susceptible to negative interference Table 1 and Table 2 show the most common chromogenic substrates for Peroxidase and Alkaline Phosphatase substrates, respectively.

Recent Advances in Immunoassays

Table 1. Common Colorimetric Peroxidase Substrates

Name	Structure	Solubility	End Product
2,2'-Azino-bis(3-ethylbenzothiazoline-6-sulfonic acid) diammonium salt (ABTS)		Soluble in water	Soluble end product that is green in color and can be read at 405 nm. The reaction may be stopped with 1% sodium dodecyl sulfate (SDS).
5-Aminosalicylic acid (5-ASA)		Soluble in water	brown in color and can be read at 450 nm. The reaction may be stopped with 3 N NaOH and read at 550 nm.
4-Chloronaphthol (4-CN)		Soluble in methanol	Used for immunoblotting, western blots
3,3'-Diaminobenzidine tetrahydrochloride (DAB)		Soluble in water	Insoluble end product brown in color and can be used as a imunohistology subdrate
o-Dianisidine dihydrochloride (DAD)		Soluble in water	Soluble end product that is yellow-orange in color and can be read at 405 nm. The reaction may be stopped with 5 M HCl.
o-Phenylenediamine (OPD)		Soluble in water	Soluble end product that is orange-brown in color and can be read at 450 nm. The OPD reaction may be stopped with 3 N HCl or 3 M H_2SO_4 and read at 492 nm.
3,3',5,5'-Tetramethylbenzidine dihydrochloride (TMB)		Soluble in water	Soluble end product that is pale blue in color and can be read at 370 or 620-650 nm. The TMB reaction may be stopped with 2 M H_2SO_4 (resulting in a yellow color), and read at 450 nm.

Recent Advances in Immunoassays

Table 2. Common Colorimetric Phosphatase Substrates

Name	Structure	Vendor	End Product
p-Nitrophenylphosphate (pNPP) Substrate		Sigma-Aldrich, USA or others	Develops a soluble yellow reaction product that may be read at 405 nm. For endpoint assays, the reaction may be stopped with 3 N NaOH. Suitable for ELISA
BluePhos® Phosphatase Substrate. Properiority formulation using modified form of 5-bromo-4-chloro-3-indolyl phosphate (BCIP) and nitroblue tetrazolium (NBT)		KPL Inc, USA	Develops a deep blue reaction The substrate remains blue after stopping and is read at 595-650 nm; perrmits accurate, quantitative measurement in kinetic ELISA. The end product produces a strong color for direct visualization.
FirePhos™ Phosphatase Substrate modified form of 5-bromo-4-chloro-3-indolyl phosphate (BCIP)		KPL Inc, USA	The end product is a red color that can be read at 500 nm after stopping the reaction.
5-bromo-4-chloro-3-indolyl phosphate BCIP) and nitroblue tetrazolium (NBT)		Thermo Fisher, USA Sigma-Aldrich, USA or Others	The combination of NBT (nitro-blue tetrazolium chloride) and BCIP (5-bromo-4-chloro-3'-indolyphosphate p-toluidine salt) yields an intense, insoluble black-purple precipitate when reacted with alkaline phosphatase, a popular enzyme conjugate for antibody probes. Suitable for immunoblotting or western blots

Recent Advances in Immunoassays

The chemical reaction of BCIP/NBT substrates with Alkaline Phosphatase is shown in Figure 4. BCIP is hydrolyzed by alkaline phosphatase to form an intermediate that undergoes dimerization to produce an indigo dye. The NBT is reduced to the NBT-formazan by the two reducing equivalents generated by the dimerization. This reaction proceeds at a steady rate, allowing accurate control of the relative sensitivity and control of the development of the reaction

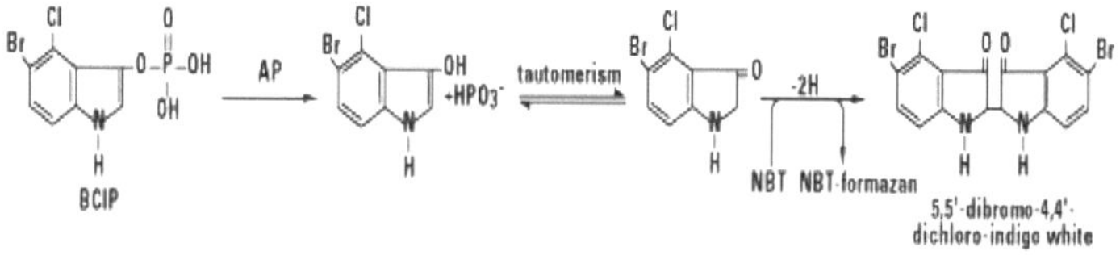

Figure 4. Chemical Reaction of BCIP/ NBT Substrate with Alkaline Phosphatase.

6. Fluorescent Immunoassay

Fluorescent immunoassay (FIA) is analogous to absorbance assay except that the label is a fluorophore rather than a chromogenic substrate. As in other immunoassays, FIA can be categorized into heterogeneous and

Recent Advances in Immunoassays

homogeneous assays, depending on whether the separation step is or is not needed, respectively. Both heterogeneous and homogeneous assays can be performed in a competitive or non-competitive format. Heterogeneous competitive FIA methods are currently being used in the analysis of various pharmaceutical compounds with the sensitivities ranged from 0.01-2.0 ng/mL. In homogeneous FIA, the antibody-bound analyte is not needed to be separated from the free analyte before the fluorescence measurement. Almost all the homogeneous FIA applied in pharmaceutical analysis are performed as competitive. In these assays, the antibody binding causes some changes in the fluorescence properties (e.g. polarization) of the labeled analyte. The analyte concentration in a sample can be monitored directly from the reaction mixture. The most common type of these assays is the fluorescence polarization fluorescent immunoassay (FPFIA). This technique is based on the following: when a fluorescent analyte conjugate is excited with polarized light, the

Recent Advances in Immunoassays

polarization of the resulting emission depends inversely on the decay constant of the probe (4-5 ns for fluorescein isothiocyanate) and on the rotational motion of the conjugate. With small molecules (e.g. drugs), random rotation decreases the polarization signal; when bound to specific antibodies, their rotation slows, and the polarization signal increases. The increase in the polarization signal is related to the concentration of the analyte. The FPFIA methods utilizing automated analyzer are simple, precise, and easy to perform.

Over 50 FP assays are commercially available for detecting various molecules including proteins (34, 35), nucleic acids (36), serum antibodies (37), and enzyme activity (38-40). However, as with any technology, there are advantages and disadvantages to FP assys. A key advantage is that minimal sample preparation is required. An FP assay can be carried out in any matrix in which the probe and antigen can interact. Analyses have been conducted in phosphate-buffered saline (PBS), sera, milk, and other solvents without any

Recent Advances in Immunoassays

sample processing (37, 41). Because of the limited need for sample processing, FP antibody detection assays are particularly useful for high-throughput screening. A key disadvantage is the limited number of small molecules suitable as probes. However, suitable probes may be developed from recombinant antibodies, thereby expanding the array of potential assays and can widely applied in drug discovery studies as well as in therapeutic monitoring of a wide range of analytes.

Fluorescent Immunoassay is much more sensitive than absorbance because the emission light signal, although weak, is measured against a background that is dark, except for scatter from the excitation source. However, fluorescence is very susceptible to negative interference (quenching) and positive interference (from native fluorescence of serum constituents and contaminants). Direct fluorescent labeling has been successfully automated using dissociation enhanced lanthanide fluorescence immunoassay

(9-12).In this system, the label is a chelate of a lanthanide ion, such as europium (Eu3+). The chelate is non-fluorescent, but the Eu3+ becomes fluorescent on addition of a dissociating reagent containing an enhancer. The dissociation allows fluorescence to be measured under optimum solution conditions, and time resolved fluorometry distinguishes the very long-lived Eu3+ fluorescence from that of contaminants (22).

7. Chemiluminescent Immunoassay

Chemiluminescence is the most intrinsically sensitive optical signal available. Photons of light arise spontaneously in the course of a chemical reaction and, therefore, can be measured against an absolutely dark background. There is little interfering chemiluminescence from serum components and contaminants. However, the label must be made to undergo a chemical reaction for the luminescence to occur. If there is any tendency for this reaction to occur spontaneously, background noise will be produced. The signal is time-dependent, which increases the difficulty of

Recent Advances in Immunoassays

reproducible measurement. Also, like fluorescence, chemiluminescence can be modulated by the solution environment. Luminol derivatives and, acridinium esters have been used commonly for direct chemiluminescent labeling in immunoassay (42-44). These molecules are relatively stable under ambient conditions, but on addition of an oxidizing reagent, they become luminescent. The common fluorescent and chemiluminescent substrates for peroxidase and alkaline phosphatase are shown in Table 3 and Table 4, respectively.

Recent Advances in Immunoassays

Table 3. Common Fluorescent and Chemiluminescent Peroxidase Substrates

Name	Stucture	Emax / Amax (Emission/Excitation	Vendor	End Product
QuantaBlu™ Fluorogenic Substrate	Properitory Information	Emax 420 nm AMax 325 nm	Thermo Scientific, USA	A larger linear detection range with low-end linearity for detection of HRP. The stable fluorescent reaction product has an Emax/Amax of 420 nm/325 nm allowing stopped, non-stopped and kinetic assays to be performed
Amplex Red® (10-Acetyl-3,7- dihydroxy-phenoxazine). Also sold as as QuantaRed Fluorogenic Substrate		Emax585 nm Amax/ 570 nm	Invitrogen, USA Thermo Scientific, USA	The fluorescent reaction product (resorufin) is stable for 4 hours with an Emax/Amax of 585 nm/570 nm when the reaction is stopped. The red-shifted resorufin reaction product permits detection at a wavelength less interference from autofluorescence that can occur in biological samples.
Luminol (5-Amino-2,3-dihydro-1,4-phthalazinedione) chemiluminescent Substrate		Emax 425 nm	Sigama-Alrich, USA	Used for Chemilumiscen substrates
SuperSignal Chemiluminescent Substrate	Properitory Information	Emax 425 nm	Thermo Scientific	SuperSignal chemiluminescent substratesis one of the most sensitive substrates available for ELISA applications. When properly optimized, the lower detection limit is 1 to 10 orders of magnitude lower than commonly used colorimetric substrates.
LumiGLO® Chemiluminescent Substrate	Properitory Information	Emax 428 nm	KPL Inc	LumiGLO is a luminol-based substrate that enables picogram-level detection of peroxidase conjugates. After reaction with HRP conjugate, light emission reaches maximum intensity within 5 minutes and is sustained for approximately 1 - 2 hours.
Novex® ECL Chemiluminescent Substrate	Luminol plus an Enhancer (properitory)	Emax 461-466 nm	Invitrogen, USA	Novex® ECL is a Chemiluminescent substrate used for chemiluminescence-based immunodetection of horse radish peroxidase (HRP) on western or dot blot membranes

Recent Advances in Immunoassays

Table 4. Common Fluorescent and Chemiluminescent Alkaline Phosphatase Substrates

Name	Structure	Vendor	End Product
p-Nitrophenylphosphate (pNPP) Substrate		Sigma-Aldrich, USA or others	Develops a soluble yellow reaction product that may be read at 405 nm. For endpoint assays, the reaction may be stopped with 3 N NaOH. Suitable for ELISA
BluePhos® Phosphatase Substrate. Properiority formulation using modified form of 5-bromo-4-chloro-3-indolyl phosphate (BCIP) and nitroblue tetrazolium (NBT)		KPL Inc, USA	Develops a deep blue reaction The substrate remains blue after stopping and is read at 595-650 nm; permits accurate, quantitative measurement in kinetic ELISA. The end product produces a strong color for direct visualization.
FirePhos™ Phosphatase Substrate modified form of 5-bromo-4-chloro-3-indolyl phosphate (BCIP)		KPL Inc, USA	The end product is a red color that can be read at 500 nm after stopping the reaction.
5-bromo-4-chloro-3-indolyl phosphate BCIP) and nitroblue tetrazolium (NBT)		Thermo Fisher, USA Sigma-Aldrich, USA or Others	The combination of NBT (nitro-blue tetrazolium chloride) and BCIP (5-bromo-4-chloro-3'-indolyphosphate p-toluidine salt) yields an intense, insoluble black-purple precipitate when reacted with alkaline phosphatase, a popular enzyme conjugate for antibody probes. Suitable for immunoblotting or western blots

Recent Advances in Immunoassays

Recently, chemiluminescent Alkaline Phosphatase ELISA detection system using CSPD® or CDP-Star® 1,2-dioxetane substrates for alkaline phosphatase with Sapphire-II™ or Emerald–II™enhancer in a system designed for rapid and ultrasensitive analyte detection in enzyme-linked immunoassays are easily available (45–51). Maximum light emission is reached in 5 to 60 minutes, depending on the temperature and the substrate chosen. Light emission can be quantitated with a variety of luminometers without the need for solution injection. The high sensitivity obtained with 1, 2-dioxetane substrates is demonstrated in a sandwich immunoassay format ELISA that employs a biotinylated detector antibody and streptavidin alkaline phosphatase conjugate for quantitating recombinant human IL-6. The results obtained with CSPD® substrate/Sapphire-II™ enhancer (Figure 1) show a significant improvement in signal-to-noise performance at all concentrations of IL-6 and a much wider assay dynamic range compared to those obtained with the fluorescent substrate 4-

Recent Advances in Immunoassays

methylumbelliferyl phosphate (4-MUP), and the colorimetric substrate, *p*-nitrophenylphosphate (*p*NPP).

Sandwich immunoassay formats with 1,2-dioxetane substrates have been used for the quantitation of a variety of animal and human proteins from plasma and tissue extracts. (52).

CDP–*Star*® substrate with Sapphire-II™ enhancer or Emerald-II™ enhancer has become widely used for immunoassay protein detection applications such as detection of plasma proteins (52) and viral antigens in both clinical serum samples (53) and in cell culture media as a viral infection. Competitive ELISA formats with 1, 2-dioxetane substrates have been utilized in these assays for detection of peptides and hormones (52, 53). The mechanism of phosphatase-dependent chemiluminescence generation by CDP-*Star* substrate is shown in Figure 5. Luminescence emission is shifted to 461 nm or 542 nm respectively by the Sapphire-II or Emerald-II enhancers.

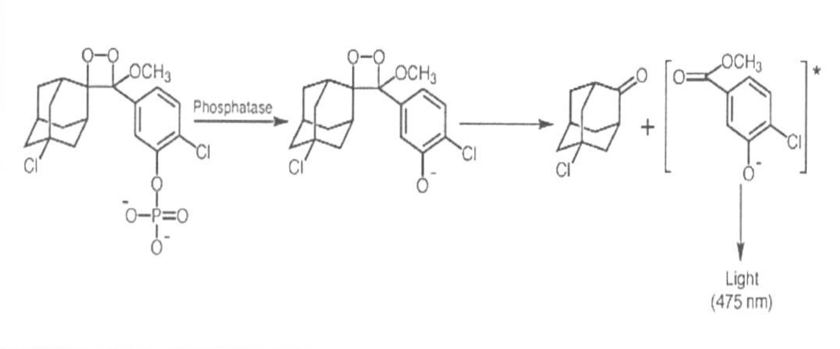

Recent Advances in Immunoassays

Figure 5. Mechanism of phosphatase-dependent chemiluminescence generation by CDP-Star®substrate.

Sensitivity can be enhanced even further by using enzyme amplification. Suitable substrates include adamantyl-1,2-dioxetane phosphates that, when hydrolyzed by alkaline phosphatase, become unstable and luminesce with relatively long decay time (53).

PerkinElmer (Waltham, Massachusetts) introduced a homogeneous immunoassay AlphaLISA® and AlphaScreen® Assays alternative to classical ELISA. AlphaLISA assays were originally utilized to detect analytes in cell cultures supernatants, or serum/plasma samples. More recently, AlphaLISA has been applied to a wider variety of biological matrices including lysates from cultured cells, or

32

Recent Advances in Immunoassays

fluids and tissue homogenates from animals. AlphaLISA assay have been used in the quantitation of cytokines and a variety of biomarkers (53-55).

AlphaLISA and AlphaScreen® technology allows the detection of molecules of interest in buffer, cell culture media, serum and plasma in a highly sensitive, quantitative, reproducible and user-friendly mode. In an AlphaLISA assay, a Biotinylated Anti-Analyte Antibody binds to the Streptavidin-coated Donor beads while another Anti-Analyte antibody is conjugated to AlphaLISA Acceptor beads or AlphaScreen® Acceptor beads. In the presence of the analyte, the beads come into close proximity. The excitation of the Donor beads provokes the release of singlet oxygen molecules that triggers a cascade of energy transfer in the Acceptor beads, resulting in a sharp peak of light emission at 615 nm (see Figure 6). For AlphaScreen® assay the light emission is 520-620 nm. AlphaScreen and AlphaLISA assays can be measured on any PerkinElmer Alpha-enabled

Recent Advances in Immunoassays

plate reader equipped with a dedicated laser for excitation, ambient temperature control and HTS apertures designed to measure the signal straight above the well.

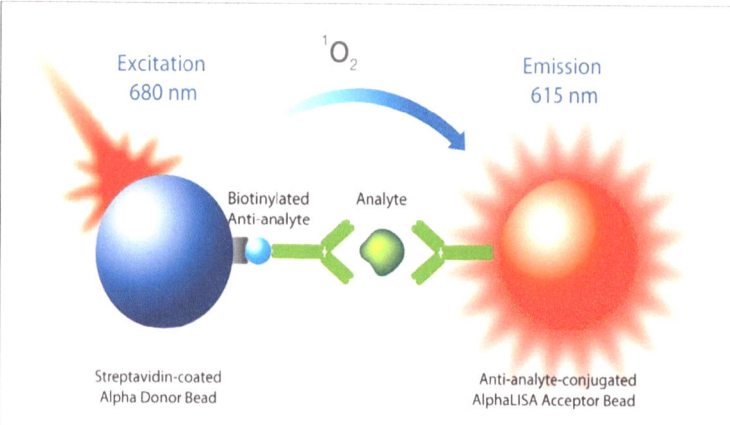

Figure 6. The Principle of the AlphaLISA Assay (©2005-2014, PerkinElmer, Inc. All rights reserved. Printed with permission).

Singulex (Almeda, CA) has pioneered digital single-molecule counting (SMC) technology that can quantitatively measure proteins and metabolites directly in complex biological samples. Proprietary digital detection technology and high precision immunoassays are combined in the Erenna® Immunoassay System to provide higher

Recent Advances in Immunoassays

sensitivity and broader dynamic range over traditional immunoassay platforms. The Erenna®

System utilizes a bead suspension sandwich assay to directly capture the analyte, and the eluate is then read on a separate detection plate. This unique assay approach, along with Singulex's proprietary Sgx link™ software algorithm, provides a lower background and increased slope, enabling quantification of low abundance biomarkers and monitoring of levels over time. In a direct comparison with a traditional ELISA-based assay, Singulex single-molecule counting technology shows improved assay sensitivity. This head to head comparison demonstrates significant improvement in detection capabilities by migrating it to the Singulex high precision digital immunoassay platform (56-59). Singulex's patented ultra-sensitive immunoassay technology enhances the clinical utility of biomarkers implicated for cardiac dysfunction, vascular inflammation and dysfunction, dyslipidemia, and cardio metabolic disorders, which provides greater

Recent Advances in Immunoassays

resolution to test results and supports better patient care (58) see at: http://www.singulex.com/ cardiovascular-health-management

and comprehensive test menu features on Cardiac Troponin-I (cTnI), the most sensitive cardiac troponin-I (cTnI) test available, as well as our proprietary inflammatory test panel including IL-6, IL-17A, TNF-α, and VEGF. Figure 7 describes the principle of Erenna® Immunoassay.

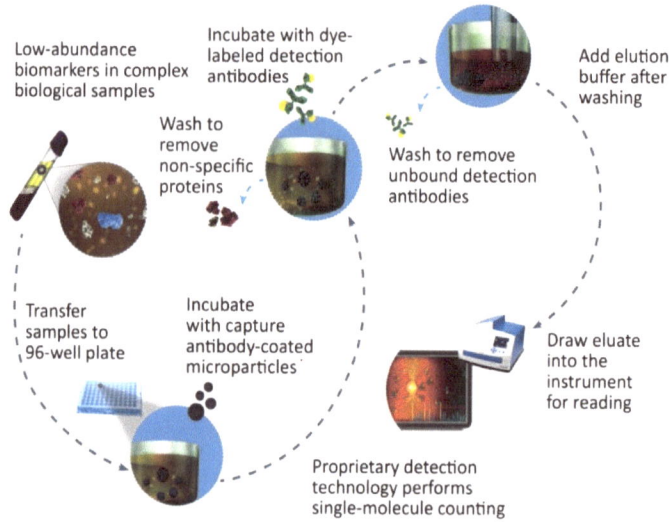

Figure 7.Principle of the Singlulex Erenna® Immunoassay Technology (Copyright Permission obtained from Singulex)

Recent Advances in Immunoassays

8. Electrochemiluminescence

Electrochemiluminescence (ECL) is chemilumilu-
nescence arising as a result of electrochemical
reactions. It includes electrochemical initiation of
ordinary chemiluminescent reactions, electrochemical modification of an analyte enabling it to
take part in a chemiluminescent reaction, or
electron transfer reactions between radicals or
ions generated at electrodes. Prominent in the
work done on electrochemiluminescence are
reactions involving polyaromatic hydrocarbons or
transition metal complexes, especially those of
ruthenium, palladium, osmium and platinum.

Applications have made use of the sensitivity,
selectivity and wide working range of analytical
chemiluminescence, but electrochemiluminescence offers additional advantages without adding
much to the inexpensive instrumentation (60).
Electrodes can be designed to achieve maximum
detection of the light emitted and electrochemical
measurements can be made simultaneously with
the light output. Generation of chemiluminescence

Recent Advances in Immunoassays

reagents at electrodes gives control over the course of light producing reactions, which can effectively be switched on and off by alteration of the applied potential; this is particularly useful when using unstable reagents or intermediates. Other possible benefits include generation of reagents from inactive precursors and regeneration of reagents, which permits the use of lower concentrations or immobilization of the reagents on the electrode. Analytes can also be regenerated, so that each analyte molecule can produce many photons, increasing sensitivity, or they can be modified to make them detectable by the chemiluminescence reaction in use. Electrochemiluminescence can be coupled with high performance liquid chromatography or with capillary electrophoresis.

The use of *tris*-(2,2$^{/}$-bipyridyl)ruthenium(II) [Ru(bipy)$_3$]$^{2+}$ in electrochemiluminescence rests on its activity with very high efficiency at easily accessible potentials and ambient temperature in aqueous buffer solutions in the presence of

Recent Advances in Immunoassays

dissolved oxygen and other impurities (61). The reaction sequence that leads to electrochemiluminescence is shown in equations in Figure 8.

(1) Oxidation: $[Ru(bipy)_3]^{2+} - e^- \rightarrow [Ru(bipy)_3]^{3+}$

(2) Reduction by analyte: $[Ru(bipy)_3]2+ + e- \rightarrow [Ru(bipy)_3]+$

(3) Electron transfer: $[Ru(bipy)_3]3+ + [Ru(bipy)_3]+ \rightarrow [Ru(bipy)_3]2+ + [Ru(bipy)_3]2+^*$

(4) Chemiluminescence: $[R(bipy)_3]2+^* \rightarrow [Ru(bipy)_3]2+ + light$

Figure 8.Reaction Sequence in the
Electrochemiluminescence

The oxidation occurs electrochemically at the anode, whereas the reduction is brought about chemically by the analyte in the free solution. Electron transfer and subsequent chemiluminescence also occur in the free solution close to the anode, where the $[Ru(bipy)_3]^{3+}$ is concentrated. Other analytes, e.g. alkyl amines, are oxidized at the anode to form a highly reducing radical intermediate that reacts with $[Ru(bipy)_3]^{3+}$ to form $[Ru(bipy)_3]^{2+*}$, which emits light. Oxalates, on the other hand, are oxidized by $[Ru(bipy)_3]^{3+}$ to

Recent Advances in Immunoassays

radicals that then reduce more $[Ru(bipy)_3]^{3+}$ to give $[Ru(bipy)_3]^{2+*}$ and chemiluminescence. ECL proved to be very useful in analytical applications as a highly sensitive and selective method. It combines analytical advantages of chemiluminescent analysis (absence of background optical signal) with ease of reaction control by applying electrode potential. As an analytical technique it presents outstanding advantages over other common analytical methods due to its versatility, simplified optical setup compared with photoluminescence (PL), and good temporal and spatial control compared with chemiluminescence (CL). Enhanced selectivity of ECL analysis is reached by variation of electrode potential thus controlling species that are oxidized/reduced at the electrode and take part in ECL reaction.

One such system, ORIGEN®, (IGEN International, Gaithersburg, MD, USA) uses ECL technology. Besides the ORIGEN system, ECL technology has been incorporated into the Elecsys®

Recent Advances in Immunoassays

Immunoanalyzer Roche Diagnostics, Indianapolis, IN), NucliSens® amplification technology (Organon Teknika, Durham, NC), and the QPCR® System 5000 (PerkinElmer, Wellesley, MA). ECL system results in an amplified signal and requires only a small volume of reagent per test. The magnetic beads provide a greater surface area than that of conventional ELISA, so the reaction does not suffer from the same surface steric and diffusion limitations. Instead, it occurs in a turbulent bead suspension, allowing for rapid reaction kinetics and a short incubation time. Detection limits of 200 fmol/L are feasible with a linear dynamic range spanning six orders of magnitude (62).

The ECL system has been demonstrated to be effective for detecting both toxins and infectious disease (ID) agents (63, 64) and could potentially be used with any biological agent, as long as high-quality, high-affinity antibodies or other ligands to those agents are available. While in general, ECL

Recent Advances in Immunoassays

assays are simple, rapid, and sensitive, assay sensitivities may vary significantly depending on the sample matrices encountered. Because of this, matrix-specific positive and negative control samples are used to establish standard curves and cutoff values. The major limitations of ECL assays are associated with the instrumentation itself and the time required to analyze each assay tube.

Meso Scale Discovery's technology (MSD, Rockville, Maryland) developed a proprietary MULTI-ARRAY® and MULTI-SPOT® microplates with electrodes integrated into the bottom of the plate. MSD's electrodes are made from carbon, an excellent material for biological assays. Biological reagents can be attached to the carbon simply by passive adsorption and retain a high level of biological activity. MSD assays use electrochemiluminescent labels for ultra-sensitive detection. These labels are non-radioactive, stable and offer a choice of convenient coupling chemistries. Electrochemiluminescent labels emit

Recent Advances in Immunoassays

light when electrochemically stimulated. The detection process is initiated at electrodes located in the bottom of MSD's microplates. Only labels near the electrode are excited and detected, enabling non-washed assays. MSD's labels emit light at 620 nm, eliminating problems with color quenching. Few compounds interfere with the electrochemiluminescence process so you can use large, diverse libraries with confidence. Multiple excitation cycles of each label amplify the signal to enhance light levels and improve sensitivity. Background signals are minimal because the stimulation mechanism (electricity) is decoupled from the signal (light). Electrochemiluminescence detection offers a unique combination of sensitivity, dynamic range and convenience. Arrays bring speed and high density of information to discovery through miniaturization, organization and parallel processing of biological assays. Proprietary electronics and efficient signal processing algorithms convert the measured signal into useful

Recent Advances in Immunoassays

data quickly, keeping read time fast, even for high-density plate formats.

Meso Scale Discovery's MULTI-ARRAY® Technology is a multiplex immunoassay system that enables the measurement of biomarkers utilizing the next generation of electrochemiluminescent detection. In an MSD® assay, specific Capture Antibodies for the analytes are coated in arrays in each well of a 96-well carbon electrode plate surface. The detection system uses patented SULFO-TAG™ labels, which emit light upon electrochemical stimulation initiated at the electrode surfaces of the MULTI-ARRAY and MULTI-SPOT® plates. The electrical stimulation is decoupled from the output signal, which is light, to generate assays with minimal background. MSD labels can be conveniently conjugated to biological molecules, are stable and are non-radioactive. Additionally, only labels near the electrode surface are detected, enabling non-washed assays.

Recent Advances in Immunoassays

One of the advantages of MSD assays is the minimal sample volume required as compared to a traditional ELISA, which is also limited by its inability to measure more than a single analyte. With an MSD assay, ten different biomarkers can be analyzed simultaneously using as little as 10-25 µL of sample. These assays have high sensitivity, up to five logs of linear dynamic range, and excellent performance in complex biological matrices. Combined, these advantages enable the measurement of native levels of biomarkers in normal and diseased samples without multiple dilutions. Further, the simple and rapid protocols of MSD assays provide a powerful tool to generate reproducible and reliable results. The MSD product line offers a diverse menu of assay kits for profiling biomarkers, cell signaling pathways, and other applications, as well as a variety of plates and reagents for assay development. The assays employ a sandwich immunoassay format where capture antibodies are coated in a single spot, or in a patterned array, on the bottom of the wells of a MULTIARRAY or MULTI-SPOT plate in 1-spot

Recent Advances in Immunoassays

MULTI-ARRAY and 4-, 7-, and 10-spot MULTI-SPOT 96-well plate formats. The assay follows a typical sandwich assay format and the intensity of emitted light is then measured in a MSD instrument to quantitate the sample (65-70). MSD multiplex assays for cytokines exhibit high sensitivity (1-10 pg/mL) and dynamic range >3 logs which allows user to avoid dilution of samples and simultaneous detection of multiple cytokines in multi spot wells.

9. Lateral Flow immunoassay

Lateral flow immunoassays (LFIA) also known as lateral flow immunochromatographic assays, are simple devices intended to detect the presence (or absence) of a target analyte in sample (matrix) without the need for specialized and costly equipment, though many lab based applications exist that are supported by reading equipment or by naked eye. Lateral flow assays have been a popular platform for diagnostic tests since their introduction in the late 1980s. Lateral flow tests are used for the specific qualitative or semi-

Recent Advances in Immunoassays

quantitative detection of many analytes including antigens, antibodies, and even the products of nucleic acid amplification tests. Also known as "hand-held" assays (HHA), they are simple to use and require minimal training. In most cases, the manufacturer provides simple instructions that include pictures of positive and negative results. Typically, these tests are used for medical diagnostics either for home testing, point of care testing, or laboratory use (71- 73).

The basic component of LFIA is composed of the following components. These are:

1. Sample Pad
2. Conjugate Pad- The conjugate pad contains detection particles (conjugate) adsorbed with antibodies or antigens specific to the analyte being detected
3. Detection Conjugate
4. Solid-phase Membrane
5. Test and control reagent lines
6. Absorbent Pad
7. Plastic-adhesive backing card

Recent Advances in Immunoassays

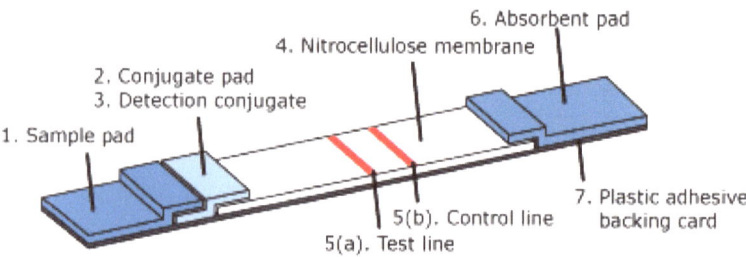

Figure 9. Principle of Lateral Flow Immunoassays

The method used for determining may a competitive or an antigen-capture assay format. The device contains a reporter antibody labeled with a colored particle such as colloidal gold, selenium, carbon, paramagnetic, or colored latex beads, which is deposited in a reservoir pad. An analyte-specific capture antibody is immobilized on the membrane. When the sample is placed on the sample application pad, the solution enters the reservoir pad and solubilizes the labeled reporter antibody, which binds to the target analyte. This analyte–antibody complex flows with the liquid sample laterally along the surface of the strip. When the complex passes over the zone where the capture antibody has been

Recent Advances in Immunoassays

immobilized, the complex binds to the capture antibody and is trapped, accumulating and producing the appearance of a colored band at the capture zone on the strip. If the result is negative and no analyte is present in the test solution, only the control band appears in the result window. This band indicates that the liquid flowed properly up the strip. If the result is positive, two bands appear in the result window. A lateral flowstrip test can provide a yes/no determination of the presence of the target analyte or a threshold (semi-quantitative) result, typically in 5–10 min.

Lateral flow immunoassays represent a well-established and very appropriate technology when applied to a wide variety of point-of-care (POC) or field use applications. The advantages of the lateral flow immunoassay system (LFIA) are well known:

1. Ease of use- minimum operator-dependency
2. Ease to manufacture- process and equipment available

Recent Advances in Immunoassays

3. Long self-lives at room temperature for use testing of diseases, such as malaria, HIV, TB, Chlamydia etc. in third-world countries
4. Easily scalable to high-volume production
5. Use of small volume of samples
6. Can be integrated with on board electronics and reader systems

However, there are number of disadvantages which must be resolved to use LFIA as a more definitive result. The major issues are:

1. Patent situation limits its application
2. Sensitivity and Specificity issues in some cases
3. Cross contamination
4. Integration with onboard electronics and built-in QC functions challenging
5. Test-to-test reproducibility – limits applications in quantitative analysis

Recent Advances in Immunoassays

10. Point of Care

Point of Care (POC) testing is defined as "testing performed at the patient's side." The testing site is usually close to the patient such as at the bedside, the physician's clinic office, intensive care unit, emergency room, operating room, or wherever medical care may be necessary. The physician, nurse, or other health care professional attending the patient may obtain the specimen, perform the analysis, and record the test result. The main objective of POC testing is to minimize the interval between testing to diagnosis and treatment. POC testing has a range of complexity and procedures that vary from manual methodologies to automated analyzers. POC testing devices are often 'hand held' or may be small portable analyzers. POC testing is generally more expensive than in lab testing but is appropriate and cost effective in some clinical settings because testing is performed near the patient and informs immediate decisions for clinical management of the patient. Microfluidics is an attractive technology for point-of-care

Recent Advances in Immunoassays

immunoassays. Heterogeneous immunoassays, because they involve capture of analytes at surfaces, are well-suited to exploit the large surface-to-volume ratios encountered in microfluidics. Faster analysis times can be achieved because of the replenishment of analytes and detection reagents in the boundary layer above the surface in standard well plate formats.

Critical among these advantages are the POC nature and a very broad range of applications that can be brought to market extremely quickly and for a relatively small investment. These are advantages that few other putative POC technologies currently in development, including sensor- and array based technologies, can claim to share. While innovation in microfluidics, biosensor, and multiplexed arrays continues at an increasing rate, those technologies typically require long development cycles, careful market selection, market education, and large investment in technology and infrastructure development in

Recent Advances in Immunoassays

order to make significant impacts in most diagnostic marketplaces.

Recently, several approaches are being explored to retain the simplicity of the HHA format while incorporating quantitative detection and improved sensitivity. Incorporation of fluorescent microspheres into modified versions of existing lateral flow assays permits the assessment of the result by a compatible reader. One such reader, the Rapid Analyte Measurement Platform Reader or RAMP™ Reader, produced by Response Biomedical Corporation (Burnaby, BC, Canada), allows for quantitative interpretation of the lateral flow assay and has been demonstrated for clinical and biodefense applications. Up-converting phosphors have also been used to make quantitative assays and also increase sensitivity (74-76). Another promising approach uses paramagnetic particles as the label, with the magnetic flux sensed within the capture zone [Magnetic Assay Reader (MAR); Quantum Design, San Diego, CA, USA]. This approach has

Recent Advances in Immunoassays

improved sensitivity by as much as several orders of magnitude over more traditional lateral flow assays, while also permitting a quantitative measurement of antigen. To read in details about POC testing, refer to (71).

In this book i describe a few immerging POC testing methods that are based on immunoassay technology. LSI Medience Corporation (Tokyo. Japan) developed an easy to use, bench-top, chemiluminescent immunoassay analyzer that rapidly measures concentrations of emergency biomarkers from a single whole blood sample. It is named as PATHFAST. The cardiac markers it can measure are cardiac Troponin I (cTnl), CK-MB, Myoglobin, NTproBNP, D-Dimer and CRP. The principle of the PARHFAST assay is shown in Figure 10.

Recent Advances in Immunoassays

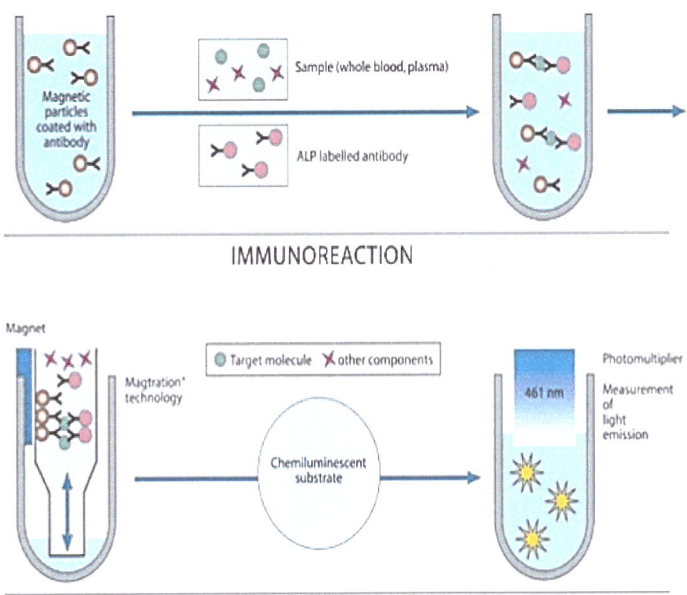

IMMUNOREACTION

SEPARATION   ENZYME REACTION   DETECTION

Figure 10. Principle of PATHFAST Fluorescent Immunoassay (copy right permission obtained from LSI Medience Corporation)

The Alere Diagnostics (also known as Biosite, San Diego, California) developed a fluorescence immunoassay for use in POC using meters to improve a physician's ability to aid in the diagnosis of critical diseases and health conditions including heart failure and myocardial infarction as well as

55

Recent Advances in Immunoassays

aid in assessing patients for pulmonary embolism. Alere Triage®MeterPro, is used for the quantitative measurements of Myoglobin, CK-MB and Troponin I as an aid in the diagnosis of myocardial infarction injury. Alere Triage® brand rapid tests include quantitative BNP, CK-MB, d-dimer, myoglobin, NGAL, troponin I (cTnI). Abbott Laboratories (Abbott Park, Illinois, USA) acquired the Biosite's core fluorescence immunoassay technology, modified and sold as Abbott i-STAT system. Both Biosite and Abbott systems provide accurate results. However, Abbott's i-STAT System has a superior menu comprised of tests used in the diagnosis, treatment and/or assessment of disease severity in patients in the emergency department and Critical Care areas of the hospital (77-81). Other POC testing methods that are available are GEM Immuno, Radiometer AQT90, and the Ortho-Clinical Diagnostic Vitros assays. In a study of cardiac troponin I for myocardial infraction, the results of all five POC testing showed highly variable sensitivity and specificity:

Recent Advances in Immunoassays

GEM Immuno (sensitivity 63%, specificity 85%), PATHFAST (sensitivity 53%, specificity 86%), Ortho Vitros (sensitivity 68%, specificity 81%), were better than AQT90 (sensitivity 26%, specificity 93%) and i-STAT (sensitivity 32%, specificity 92%) (82). The observed concentrations between each assay were different which is not unexpected because standard used in each assay was different from each other.

11. Time-Resolved Fluorescence

In time-resolved fluorescence (TRF), a fluorescent probe is employed that has fluorescence decay (lifetime) that substantially exceeds the duration of the exciting pulse and the duration of the background non-specific fluorescence. A time-gating is used to reduce the background fluorescence, i.e., the measurement of the fluorescence is delayed until a certain time has elapsed from the moment of excitation. The delay time is sufficiently long for the background fluorescence to have ceased. When the

Recent Advances in Immunoassays

fluorescence signal is measured (after the delay) the measurement is an integrated measurement, i.e. all the light arriving at the detector during the measuring period is measured without regard to the time of arrival. The purpose of this delayed measurement is to ensure that only one fluorescence signal reaches the detector during measurement.

Time-resolved fluorescence (TRF), are sandwich type assays similar to those used for ELISA and ECL, except that the detector antibodies are directly labeled with lanthanide chelates such as europium, samarium, terbium, and dysprosium. Its strengths derive from its sensitivity, a similarity shared with the commonly used ELISA techniques, and the potential for multiplexing. This immunodiagnostic technology is used to detect agent-specific antibodies, microorganisms, drugs, and therapeutic agents (83, 84). TRF assays TRF exploits the differential fluorescence life span of lanthanide chelate labels compared to background fluorescence. The fluorescence signal is long-lived

Recent Advances in Immunoassays

and results in assays with a very high signal-to-noise ratio and excellent sensitivity (85). TRF produces its signal through the excitation of the lanthanide chelate by a specific wavelength of light (86).

PerkinElmer's DELFIA® (dissociation-enhanced lanthanide fluorescence immunoassay) is a time-resolved fluorescence (TRF) intensity technology. Assays are designed to detect the presence of a compound or bimolecular using lanthanide chelate labeled reagents, separating unbound reagent using wash steps. The technology is based on fluorescence of lanthanide chelates (Europium, Samarium, and Terbium). The fluorescence decay time of these lanthanide chelate labels is much longer than traditional fluorophore, allowing efficient use of temporal resolution for reduction of auto fluorescent background. The large Stokes' shift (difference between excitation and emission wavelengths) and the narrow emission peaks contribute to increasing signal-to-noise ratio (84). Sensitivity is further increased because of the

Recent Advances in Immunoassays

dissociation-enhancement principle: the lanthanide chelate is dissociated and a new highly fluorescent chelate is formed into a protective micellar solution.

DELFIA lanthanide chelates require this dissociation/enhancement step for fluorescence after adding a low pH enhancement solution (86-89). Figure 11 shows DELFIA-TRF assay principle.

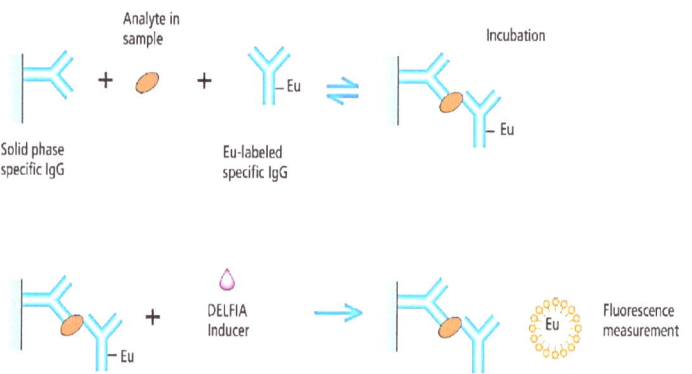

Figure 11. DELFIA TRF Sandwich type assay principle (©2005-2014, PerkinElmer, Inc. All rights reserved. Printed with permission).

[A typical agent-detection assay is similar to an ELISA except that the detector antibody is labeled directly with europium. The manufacturer claims detection limits as low as 10-17 moles of europium

Recent Advances in Immunoassays

with a dynamic range of at 4 logs and has the advantages over traditional ELISA in its application in research, drug discovery, and clinical environments].

12. Microfluidic-Based Immunoassays

Immunoassay technology based on microfluidics principles, allowing integration of the chemical operations involved in conventional analytical processes, such as mixing, reaction and separation, has been is a focus in recent years with very low volume of sample(in nanoliter scale). Gyros (www.gyros.com) uses microfluidic compact discs (Bioaffy CDs) consisting of micro columns pre-packed with biotin-labeled capture reagent, fluorophor labeled detection reagent and a complex network of microchannels with hydrophobic barriers through which samples and reagents flow under the influence of centrifugal and capillary action as the CDs are spun in the workstation with the Gyrolab™ xP workstation. By altering rotational speeds of the CD, precise volumes of capture reagents, sample and

Recent Advances in Immunoassays

detection reagents can be delivered through the micro columns. Implementation of microfluidic principles ensures controlled flow of capture reagents, sample and detection reagents, so that all structures within a single CD are processed in parallel. This eliminates the occurrence of time dependent artifacts that may be observed in a typical plate-based ELISA, and ensures samples are processed under uniform conditions. Quantification is achieved by detection of fluorescence within each microstructure using the Gyrolab Viewer software. The Bioaffy CDs are compatible with a number of conventional immunoassay formats (90).

Recently, Siloam Biosciences (www.siloambio.com) in collaboration with BioTek Instruments introduced Optimiser ™ microfluidic technology by incorporating the power of microfluidics into a standard ELISA plate with dramatic improvement in reaction efficiency and signal sensitivity in comparison with the standard assay. The microfluidic channel at the base of each well

Recent Advances in Immunoassays

serves as the "reaction chamber" thereby cutting down reaction volumes, reducing incubation times and simplifying assay operation by replacing the traditional labor-intensive wash with a simple "flush" step. The high surface area to volume ratio of the Optimiser™ allows for ultra-fast reaction kinetics translating to ~5-10 min incubation cycles perstep in an assay sequence. The Optimiser™-based ELISA use the standard ELISA workflow but uses volumes significantly lower than those used in conventional ELISAs. Moreover, there is no need to use microplate washers. As each micro channel has a volume of only ~5 µL, the addition of excesswash buffer to the well flushes out the micro channel contents onto an absorbent pad beneath the micro plate, simulating a wash process. The micro channel is arranged in a spiral pattern directly below the well of the micro plate and with the final addition of substrate, defines the detection volume. As the area of the micro channel spiral is similar to a micro plate well, a conventional fluorescence micro plate reader can be used for detection. The use of IL-4 using

Recent Advances in Immunoassays

Optimiser™ ELISA has been demonstrated in a Singleplex ELISA format (91, 92) and also in a multiplex assay with 10 analytes in activated Th17 cells (e.g., IL-6, IL17A, IL-17AF, IL-17FF, IL-21, IL-22, IL-26, G-CSF, CCL20, TNF-α) (93). Figure 12 describes the Optimiser™ ELISA operation.

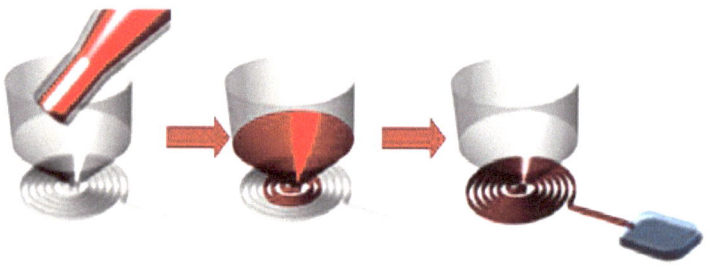

Figure 12: Principle of Optimiser™ Operation

(initial well loading with reagent serves to load micro channel. Once a well is empty and micro channel filled, capillary forces retain fluid in the micro channel for an incubation step. Addition of subsequent reagents relieves capillary forces causing flushing of initial reagent into an absorbent pad beneath the plate (shown on the side in the figure for clarity). Once the well is again empty, capillary forces induce another incubation step (Source: Copyright: © 2012 Lakkis M,

Recent Advances in Immunoassays

et al. Translational Medic 2012, S: 1 pp 1-5, 2012 an open-access article distributed under Creative Commons Attribution License)

13.   Electrochemical Immunoassay

Most immunoassay techniques are based on the separation of free and bound antigen–antibody complex. In the Electrochemical Immunoassay (often called Amperometric Immunoassay), either the antigen or the antibody is immobilized onto electrode surfaces either by adsorption or covalent attachment to form an immune complex. Quantification of bound immune complex is conducted by using labels covalently bound to the immune agent with specific properties suitable for detection. As a result, the final stage of immunoassay is in fact the quantitative detection of the label. Enzymes are the most frequently used labels linked to the antigen or to the antibody to visualize the binding event (94). Horseradish peroxidase (HRP), alkaline phosphatase (ALP) and to lesser extent ß- Galactisidase are used for enzyme amplifications. HRP has the advantage of

Recent Advances in Immunoassays

high rate for enzyme reaction and is used in electrochemical immunoassay with the use of a suitable electrochemical mediator for electrochemical detection of the catalytic reduction of hydrogen peroxide (95). However, in the practical development of immunosensors employing Horseradish peroxidase is limited due to the fact that substrate hydrogen peroxide is chemically unstable in organic reagents needed for the assay. On the other hand, alkaline phosphatase also has a high turnover rate and can be suitably used for high sensitivity electrochemical immunoassay. However, the availability of suitable phosphate substrates are limited (96). However, Mason and coworkers have developed highly sensitive electrochemical immunoassay with 4-Amino-1-naphthylphosphate (ANP) as well as 4-Hydroxynaphthyl-1-phosphate (HNP) as substrates in sandwich immunoassay (96, 97). The products of the substrates ANP and HNP after the enzyme reactions are 4-hydroxynaphthol and 4-aminonaphthol, which undergo rapid auto-oxidation to 1, 4-immino-

Recent Advances in Immunoassays

naphthaquinone and 1, 4- naphthaquinone, respectively (see Figure 13). These compounds could be easily detected in an amperometric flow injection assay (AFIA) with -200 mV versus Ag/AgCl potential application. The future development of electrochemical immunosensors and biochemical arrays based on electrochemical transducers depends on the availability of suitable phosphate substrates that can be converted by a labeling enzyme to an electrochemical active product. The electrochemically active product should also be detectable at relatively low potential with rapid electron transfer to the electrode.

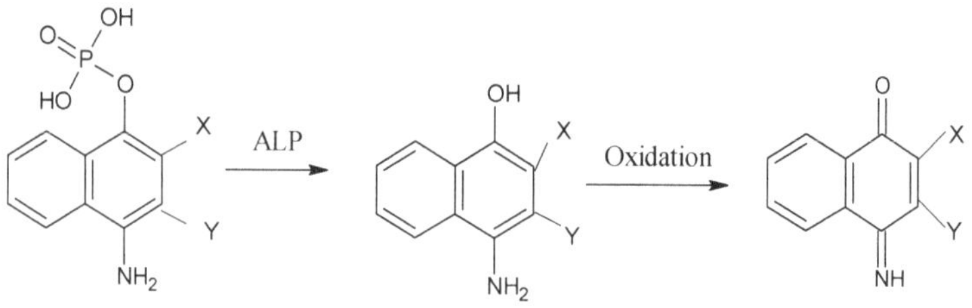

Recent Advances in Immunoassays

4-Amino-1-naphthyl-
phosphate (X, Y =H)

1, 4-Imminonaphthaquinone

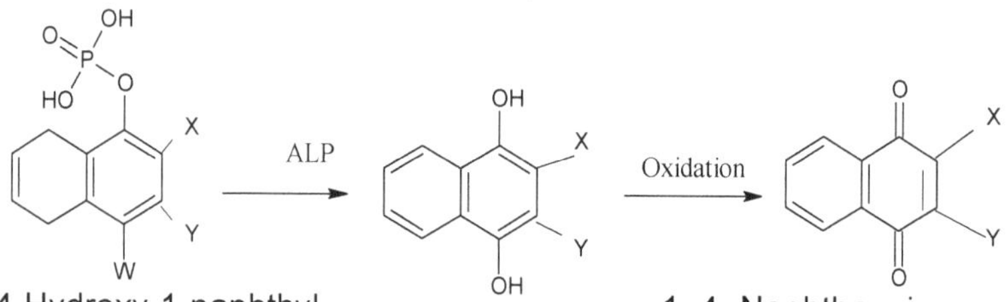

4-Hydroxy-1-naphthyl-
phosphate (X, Y =H)

1, 4- Naphthaquinone

Figure 13. Dephosphorylation of Napthylphosphate
derivatives to Quinones

14.   Multiplex Immunoassays

Multiplex assay, often called multiplex bead array assay (MBBA) is a type of assay that simultaneously measures multiple analytes (dozens or more) in a single run/cycle of the assay. It is distinguished from procedures that measure one analyte at a time. The use of

Recent Advances in Immunoassays

multiplex immunoassays for multiple analyte detection has proven to an invaluable tool for the comprehensive study of biological systems. As these systems are comprised of networks of secreted proteins including cytokines, chemokines, growth factors, and other proteins, multiplex assays allow for the biomarker profiling of a large set of proteins from a small sample. A number of multiplex assay format are known to exist, but in this book only the major ones will be covered.

Multiplex assays may be performed on either multi-use flow cytometers (such as the commonly available clinical cytometers from Becton Dickinson (BD), Beckman-Coulter, or Dako-Cytomation or on more specialized platforms such as the Luminex 100 system cytometers, increasingly MBAAs are performed on dedicated instrumentation. Three of the most widely available commercial multiplex systems are from Becton Dickinson Immunocytometry Systems, Luminex Corporation, and Diasorin. Each of these

Recent Advances in Immunoassays

will briefly be described below as examples of different approaches rather than as a comprehensive list of all manufacturers. The BD™ Cytometric Bead Array (CBA) is a flow cytometry application that allows users to quantify multiple proteins simultaneously (98-101). With BD CBA, up to 30 proteins can be analyzed using just 25 to 50 μL of sample. With the BD CBA Enhanced Sensitivity Flex Set system, it is possible to detect as low as 0.274 pg/mL in a multiplex assay.

The BD CBA system uses the broad dynamic range of fluorescence detection offered by flow cytometry and antibody-coated beads to efficiently capture analytes. The first CBA system from BD Biosciences (San Jose, CA, USA) relies on different fluorescent intensities of a single fluorophore to accomplish multiplexing. As a result, the number of assays which can be performed is more limited than in array defined by two fluorochromes (102). However, one advantage of the CBA system is that it can be performed on a clinical flow cytometer already installed in the

Recent Advances in Immunoassays

laboratory. The CBA system includes a cytometer setup kit with the requisite software, reagents and standards. The company's CBA assay kits employ their proprietary bead sets, which are internally dyed with varying intensities of a proprietary fluorophore. These sets are distinguished via one fluorescence parameter and two size discriminators. BD CBA system includes analysis software and is compatible with contemporary data acquisition software such as CellQuest™. BD has also produced the FACSArray™ Bioanalyzer, a multifunctional system that supports both the multiplexed CBA technology and limited cellular analysis. The BD FACSArray™ instrument has two lasers (green: 532 nm, red: 635 nm), detects two scatter signals (forward-scatter width and side-scatter width) and four fluorescence signals (yellow, red, far-red, and near infrared). The instrument has the capacity to use 96-well microtiter plates for samples. The company currently offers CBA kits for a wide variety of ligands, and data has been published from these kits for the quantitation of viral proteins (98) and

Recent Advances in Immunoassays

for the multiplexed detection of human cytokines (99-104). This method significantly reduces sample requirements and time to results in comparison with traditional ELISA and Western blot techniques. However, the results obtained from flow cytometry and ELISA methods do not match each other as found by Shih-Houng and coworkers on the Performance evaluation of cytometric bead assays for the measurement of lung cytokines in two rodent models (104). The sample-to-sample correlation was good between ELISA and CBA with correlation coefficient R values of 0.76, 0.66, and 0.92 for rat IFN-γ, TNF-α, and IL-6, respectively. ELISA only correlated significantly with the flow cytomix assay for TNF-α with $R = 0.43$. In conclusion, direct comparisons between absolute protein values did not agree among the assays tested in this study, but patterns of cytokine response generally agreed between ELISA and CBA. The principle of BD's CBA assay principle is shown in Figure 14.

Recent Advances in Immunoassays

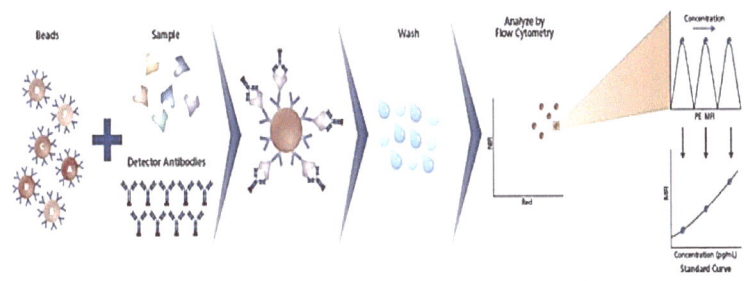

Figure 14. Principle of BD CBA Assay
(Courtesy of Becton Dickinson Biosciences)

[Capture bead in the array has unique fluorescence intensity and is coated with a capture antibody specific for a single analyte. A combination of different beads is mixed with a sample or standard and a mixture of detection antibodies that are conjugated to a reporter molecule (PE). Following incubation and subsequent washing, the samples are acquired on a flow cytometer].

The second is the Luminex xMAP technology (formerly LabMAP, FlowMetrix) that uses digital signal processing capable of classifying polystyrene beads (microspheres) dyed with distinct proportions of red and near-infrared fluorophores. These proportions define 'spectral addresses' for each bead population. As a result, up to five hundred different detection

Recent Advances in Immunoassays

reactions can be carried out simultaneously on the various bead populations in very small sample volumes (105,106).This technology is licensed by BD and is commercially available as Bio-Plex® multiplex immunoassay system. The system utilizes xMAP technology licensed from Luminex to permit the multiplexing of up to 500 different assays within a single sample. This technique involves 100 distinctly colored bead sets created by the use of two fluorescent dyes at distinct ratios. These beads can be further conjugated with a reagent specific to a particular bioassay. The reagents may include antigens, antibodies, oligonucleotides, enzyme substrates, or receptors. The technology enables multiplex immunoassays in which one antibody to a specific analyte is attached to a set of beads with the same color, and the second antibody to the analyte is attached to a fluorescent reporter dye label. The use of different colored beads enables the simultaneous multiplex detection of many other analytes in the same sample. A dual detection flow cytometer is used to sort out the different assays by bead colors in one channel

Recent Advances in Immunoassays

and determine the analyte concentration by measuring the reporter dye fluorescence in another channel. The microspheres internal dyes are excited by the laser or LED, identifying the microsphere set. A second laser or LED excites the fluorescent dye on the reporter molecule. Using these signals, high speed digital-signal processors identify each individual microsphere and quantify the amount of target molecule bound to its surface. Some recent applications with Luminex-based fluorescent microspheres have included cytokine quantitation (107), hormonal analysis (109), and biomarkers (110). The principle of Luminex xMAP technology is shown in Figure 15.

Recent Advances in Immunoassays

Figure 15. Principle of Luminex xMAP technology (copy right permission obtained from Luminex)

[Luminex xMAP technology uses color-coded tiny beads, called microspheres. Beads are colored internally with two different fluorescent dyes (red and infrared). Different concentrations of red and infrared dyes are used to generate up to 500 distinct bead regions. Each bead region is conjugated to a specific target analyte followed by binding with a biotinylated detection antibody and

Recent Advances in Immunoassays

a reporter dye, streptavidin-conjugated phycoerythrin. Inside the Luminex analyzer, a light source excites the internal dyes that identify each microsphere particle, and also the reporter dye captured during the assay. Many readings are made on each bead set, which further validates the results. Using this process, xMAP Technology allows analysis of multiple analytes in a single well]

During data acquisition, the contents of each micro plate well are drawn into the array reader, depending on the type of reader — either the flow cytometry/laser excitation–based Bio-Plex 200 and Bio-Plex 3D systems or the light emitting diode (LED)/image-based analysis employed in the Bio-Plex® MAGPIX™ multiplex reader. For the flow cytometry–based systems, precision fluidics aligns the beads in single file through a flow cell where two lasers excite the beads individually. The red classification laser excites the dyes in each bead, identifying its spectral address. The green reporter laser excites the reporter molecule associated with

Recent Advances in Immunoassays

the bead, which allows quantitation of the captured analyte. With the LED/image-based Bio-Plex MAGPIX reader, the beads are drawn into a chamber and magnetically immobilized. Classification and reporter excitation is accomplished with the use of LEDs rather than lasers. In all readers, high-speed digital signal processors and software record the fluorescent signals simultaneously for each bead, translating the signals into data for each bead-based assay (111, 112). The Bio-Plex® allows simultaneous detection of up to 500 different types of molecules in a single well of the 96-well microplate on the Bio-Plex® 3D system, up to 100 different types of molecules on the Bio-Plex® 200 system, and up to 50 different types of molecules on the Bio-Plex® MAGIPIX system The principle of Bio-Rad's flow cytometry based Bio-Plex® traditional and image based assays are illustrated in Figure16.

Recent Advances in Immunoassays

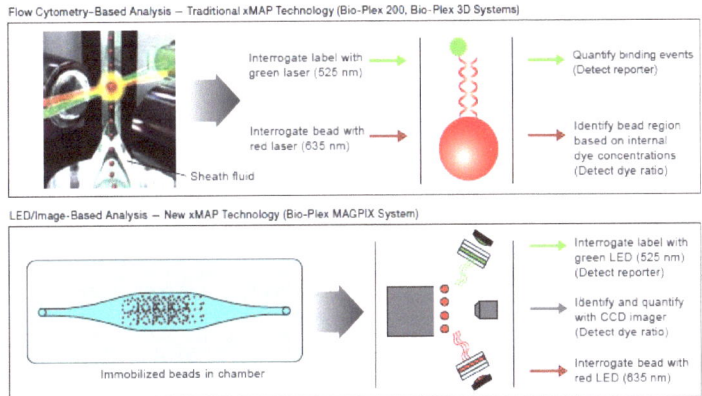

Figure 16: Bio-Rad's Bio-Plex® Traditional and Image –Based Assays (Courtesy of Bio-Rad Laboratories, Inc., © 2014)

[Dyed beads are pushed through a detection chamber in a single file or magnetically immobilized. The red classification laser (635 nm) or LED interrogates the internal dyes to identify bead regions. The green reporter laser (532 nm) or LED interrogates the fluorescent reporter to measure analyte concentration].

The third multiplex technology Copalis (coupled-particle light-scattering) developed by Diasorin, is an innovative homogeneous immunoassay technology that differs from most other multiplex

Recent Advances in Immunoassays

bead array approaches in that it does not use fluorescence to discriminate different bead populations, but rather differentiates monomeric latex microspheres from latex aggregates and cells on the basis of their unique light scatter properties by flow cytometry. The beads are differentiated on the basis of light-scatter measurements as they flow across a finely focused elliptical beam produced by a semiconductor laser. To ensure that the microparticles are centered in the flow stream, a concentric stream of sheath fluid is introduced around the sample stream. As the microparticles flow, a pulse of scattered light is produced according to their aggregation status (monomers, doublets, triplets, or larger multiplets), and this scattering is converted into a decrease in the monomer number if the reaction is positive. The system can measure two types of events: polystyrene-microparticle latex co-agglutination and polystyrene-gold colloid microparticle coupling. The former is useful for detecting the presence of antibodies to infectious agents or

Recent Advances in Immunoassays

autoantigens, which are coated onto latex microparticles. The presence of specific antibodies is detected as a reduced number of non-agglutinated particles coated with the corresponding antigen. Based on this concept, a multiplex Copalis agglutination test was developed for syphilis, Epstein-Barr viral infection, SSA/SSB autoimmunity and cytomegalovirus (CMV) (112-115). However, Copalis seems not to have been widely used – possibly due to limited multiplexing capabilities and the reduced sensitivity compared with fluorescent techniques.

15. Comparison of Multiplex Bead Arrays vs Singleplex ELISAs

The acceptance of multiplex assays for use in clinical diagnostics is largely dependent on achieving similar results to those obtained using ELISA techniques, which are widely accepted as the current 'gold standard'. Thus, a large number of studies published to date have compared data from multiplex bead assays to ELISA (103, 104, 116-127). While multiplex may be used to detect

Recent Advances in Immunoassays

and quantitate a wide assortment of ligands, much interest has centered on the quantitation of cytokines. It is difficult to compare the data from the different studies, as various investigators used different methods of comparison between multiplex assays and ELISAs from different vendors, and reports varied considerably in the amount of methodological detail provided. Although the majority of published studies have shown good correlations between multiplex assays and ELISAs but a high degree of variation is observed between these methods. These variations are likely to the result of how these comparisons were made as well as the type of antibodies used in each of the assays. Multiplex assays, by their very nature, involve potential interactions between multiple different antibodies and analytes (antigens) in the sample/assay solution. One cannot assume that a reliable ELISA can just be simply added to a functioning multiplex assay. Non-reactivity to all other antibodies is a big factor to minimize such cross-reactions. Certain proteins in biological samples, particularly

Recent Advances in Immunoassays

abundant circulating proteins in serum or plasma samples, may affect multiplex results. In the bead-based multiplex arrays, it is important to note that all reactions take place among molecules and antigens which are freely mobile in solution, while ELISA involves the immobilization of the capture antibody and thus of the resultant antigen-antibody enzyme complexes to the solid surface.

In this book, I describe some examples of the correlations between multiplex assays and Singleplex ELISAs. In a study described by Chen and coworkers (102) the capture as well as reporting antibodies were used from the same company to make both Singleplex ELISA and Cytometric Bead Array Assay with six cytokines, namely, IL-2, IL-4, IL-10, IL-12, IFN-y, and TNF-α revealed correlation coefficients ranging from 0.92 to 1.0. In another study Khan and coworkers (103) multiplex kits from four different vendors were used to measure the levels of IFN-y, IL-1ß, IL-6, IL-8, and TNF-α and compared the results with those from ELISA techniques. The results showed that the concentrations of cytokines in the multiplex assays differed for each vendor, although the cytokine levels followed similar qualitative patterns. These results clearly indicate that the antibody pairs used in the various ELISA and MBAA kits were different from each other and contribute to this variability.

Similarly, Prabhakar and coworkers (117) compared the results the cytokines between Bio-Plex Luminex assay for quantifying cytokines such as, TNF-α, IL-1ß, IL-6 and IL-8 levels in lipopolysaccharide (LPS)-stimulated human plasma samples. This study showed fairly good correlation agreement between Bio-Plex and ELISA, with correlation coefficients (r^2ranging from 0.92 to 0.98).In another study, DuPont and coworkers (119) evaluated the correlation of ELISA and Luminex MAP® techniques for quantitating a variety of cytokines in culture supernatants using kits from two commercial vendors (either from Linco and Upstate). They demonstrated good correlations between ELISA

Recent Advances in Immunoassays

and Luminex for seven cytokines (IL-1ß, IL-4, IL-5, IL-6, IL-10, IFN - y and TNF-α), fair correlations for IL-13, and a poor correlation in the case of IL-12. However, although the correlations were generally good, the authors reported a significant variation between the absolute cytokine concentrations determined by ELISA and either multiplex kit. These differences were generally more pronounced with the LINCO-Plex kit (Millipore). Similar to other investigators, DuPont and colleagues point to differences in antibody pairs and sample diluents as likely causes of observed differences. They also stress that each antibody will have optimal binding affinities at specific pH and salt concentrations, which may impact on multiplex assays (121).

In a recent, very large study of more than 2000 serum specimens, Ray et al (120) carefully examined the validation and implementation of cytokine multiplex assays, as a replacement for ELISA techniques. Although a fairly good correlation (Lin's concordance correlation =84.5%;

Recent Advances in Immunoassays

Lin's corrected for bias = 94.5%) was found between these assays, the multiplex results were, on average, 2.36-fold higher than ELISA values. The authors conclude that the difference between MBBA and ELISA assays derive from different antibody pairs and also different lots of standards as was observed by other investigators. More recently, Young and coworkers (104) compared cytokine levels of two rodent models by CBA and FlowCytomix (FC). The sample-to-sample correlation was good between ELISA and CBA with correlation coefficient R values of 0.76, 0.66, and 0.92 for rat IFN-γ, TNF-α, and IL-6, respectively. ELISA only correlated significantly with the FC assay for TNF-α with R = 0.43. For a method-to-method comparison, cytokine standards from ELISA kits were used with both ELISA and CBA to determine the R values and found it to be greater than 0.90 for all the cytokines tested. It was found that the ELISA was more sensitive in the low range of the standard curve while the bead assays were capable of detecting higher protein concentrations, which

Recent Advances in Immunoassays

would allow for direct measurement of concentrated samples. There was a lack of agreement between the absolute protein values for the ELISA and flow cytometric bead-based assays; in most cases, the latter method tended to give higher protein concentrations than ELISA. In conclusion, direct comparisons between absolute protein values did not agree among the assays tested in this study, but patterns of cytokine response generally agreed between ELISA and CBA. In the case of the mouse CBA, a companion measurement is recommended if samples with low concentrations of an analyte are reported and extrapolated below sensitivity or zero.

In a very recent study Koning and coworkers (123) compared multiplex immunoassay from Meso Scale Discovery (MSD) to Singleplex immunoassays for measuring inflammatory factors, and to examine how combining data from each affects an epidemiologic association. Plasma IL-1 ß, IFN-y, IL-6, and TNF-α were measured in 100 samples using a multiplex kit from MSD and

Recent Advances in Immunoassays

Singleplex ELISAs from R&D Systems. Separate samples (n = 80) were collected to compare multiplex and Singleplex assays from MSD. The results showed that compared to R&D ELISAs, the MSD multiplex proportionally and significantly overestimated IL-1 ß (slope = 1.2), and IFN-y (slope =2.9) but underestimated IL-6 (slope = 0.5). Correlations were ≥ 0.81 except for TNF-α (r = 0.31). Compared to MSD Singleplex, the MSD multiplex proportionally underestimated IFN-y (slope = 0.7) and TNF-α (slope = 0.5). Correlations were ≥ 0.96. These results indicate the MSD multiplex immunoassay for inflammatory factors yielded significantly different results than singleplex immunoassays—including those from the same company. Correlations were not consistently high, except among assays from the same company. Such differences may distort epidemiologic relationships if data from both methods are merged.

Recent Advances in Immunoassays

Recently a multisite comparison of high-sensitivity multiplex cytokine assays were conducted between on four different vendors (103) on Luminex (Bio-Rad, BioSource, Linco) or electrochemiluminescence (Meso Scale Discovery) platform were evaluated for their ability to detect circulating concentrations of 13 cytokines, as well as for laboratory and lot variability. Assays were performed in six different laboratories utilizing archived serum from HIV-uninfected and -infected subjects from the Multicenter AIDS Cohort Study (MACS) and the Women's Interagency uninfected and -infected subjects from the Multicenter AIDS Cohort Study (MACS) and the Women's Interagency HIV Study (WIHS) and commercial plasma samples spanning initial HIV viremia. In a majority of serum samples, interleukin-6 (IL-6), IL-8, IL-10, and TNF-α were detectable with at least three kits, while IL-1ßwas clearly detected with only one kit. No single multiplex panel detected all cytokines, and there were highly significant differences ($P < 0.001$) between laboratories and/or lots with all kits.

Recent Advances in Immunoassays

Nevertheless, the kits generally detected similar patterns of cytokine perturbation during primary HIV viremia. This multisite comparison suggests that current multiplex assays vary in their ability to measure serum and/or plasma concentrations of cytokines and may not be sufficiently reproducible for repeated determinations over a long-term study or in multiple laboratories but may be useful for longitudinal studies in which relative, rather than absolute, changes in cytokines are important.

In another study Scaife and coworkers (127) compared multiplex sandwich ELISA and bead based assays. The study showed that there was no direct comparison between the two multiplex sandwich ELISA procedures (FAST Quant and SearchLight) and a bead based assay (UpState Luminex). All three kits differed from each other for different analytes and there was no clear pattern of one system giving systematically different results than another for any analyte studied. This study clearly indicates that the results obtained from different systems cannot be combined

Recent Advances in Immunoassays

The results derived from a number of studies have demonstrated good correlations, but often poor concurrence of quantitative values, between multiplex bead array assays and corresponding Singleplex ELISA measurements. Substantial differences exist between different multiplex assays in terms of quantitative numbers (127). The comparison between multiplex assay with the gold standard Singleplex ELISA differ substantially when the reagents are different in both multiplex and Singleplex assays (103,126,127). However, the variability is minimized when the comparisons are made between both multiplex and Singleplex ELISA assays using identical capture and reporter antibodies, as well as similar diluents and serum blockers resulting in good correlation. With this in mind, any study that involves sequential monitoring of patients, or other samples, should be performed using only a single technique, one platform, and one commercial vendor for all samples otherwise. It is true that at least for cytokine multiplex assays from different vendors show similar trends in cytokine levels but the

Recent Advances in Immunoassays

absolute levels vary (126). It is therefore reasonable that the data from one platform should not be mixed. It may be possible to establish a systematic "bias" between multiplex assays when comparing the data which is a big task by itself.

Multiplxex arrays have a number of advantages when compared to Singleplex ELISA. These are:

1. High throughput multiplex analysis
2. Less sample volume needed
3. Higher dynamic range
4. Ideal for screening or profiling purpose
5. Ability to perform repeated measure of one analyte in the context of others
6. Cost effective when more than 4 cytokines are measured

Although multiplex assays provides substantial savings in assay costs, initial investment in multiplex assay equipment costs is quite high. A Luminex reader or a flow cytometer (for CBA), which range from $30,000 to more than $100,000 based on capacity. A Bio-Plex 200, for instance, costs about $50,000. As a result, at many institutions, such equipment is viewed as a shared core resource and is accessible to multiple investigators.

In contrast, singleplex ELISA plate readers can cost from between $6,000 to $20,000 depending upon their capabilities, eg. Tecan, BioTek and other instruments that measure absorbance, chemiluminescent or fluorescent signals. Similarly, a Singleplex commercial 96-well plate ELISA can cost $350−$700, while a 96-well plate multiplex assay range $600−$1,200, depending on the number of analytes being measured. Multiplexed ELISAs also require a reader; one low-cost option is Quansys' Q-View™ Imager, which costs $9,900

16.   Immunoassay based Raman Spectroscopy

Raman spectroscopy detects the scattered light off of molecules which have been illuminated by a laser. This analysis method has been used for

93

Recent Advances in Immunoassays

many years (128) to identify molecular structures in complex mixtures and is a form of vibrational spectroscopy. In principle with Raman Spectroscopy a monochromatic light source (i.e. laser) is used to illuminate a sample which reflects back a small proportion of this light as scattered light. The majority of the scattered light is of the same frequency as the excitation source. This is known as Rayleigh or elastic scattering. A very small amount of the scattered light (10^{-6} times the incident light intensity) is shifted in energy from the laser frequency. This is known as Raman scattering. Plotting the intensity of this shifted light versus frequency, results in a Raman spectrum of the sample. In such a process, each sample generates its own unique spectra based upon its molecular structure. Thus, a Raman spectrum can be used to uniquely identify a molecule.

This instrument performs these operations, using light from a 532 nm low power laser to generate and collect Raman spectra from samples contained within the designated Microtiter plate

Recent Advances in Immunoassays

wells. Information from each Raman spectra is subsequently transformed into a single numerical value, which can be correlated back to attributes of the well contents. Sword Diagnostics (Sworddiagnostics.com) evaluated the detection of Raman-active molecules that are generated in the enzyme immunoassay (ELISA) that can be measured using the fluorescence detection channel of commercially available microtiter plate readers, and is treated similarly to how absorbance, fluorescence, and/or luminescence data is processed (129, 130).

Sword Diagnostics developed a Raman detection system with commonly used "Sandwich" ELISA assays utilizing a 96 well microtiter plate bound capture antibody. The Sword Raman detection system replaces typical ELISA detection moieties (Absorbance, Fluorescence, and/or Chemi-luminescence based horseradish peroxidase (HRPO) substrates with which upon HRPO oxidation generates a large Raman signal (130). The reaction chemistry is compatible with the

Recent Advances in Immunoassays

typical ELISA reagents, and does not require the reconfiguration of standard ELISA assay (new biotinylated probes and/or conjugates are not required). After stopping this end-point reaction, the Raman signal is measured in using the fluorescence detection channel of commercially available microtiter plate readers, and is treated similarly to how absorbance, fluorescence, and/or luminescence data is processed.

The chemistry component for peroxidase detection utilizes 5-Aminosalicyclic acid and stabilized peroxide in a buffered solution at pH 6.5, which in the presence of peroxidase (horseradish) forms an iminoquinone. A Raman signal is subsequently generated from the reaction product upon the addition of sodium hydroxide development agent. The Raman-active product is then measured. This priniciple of the reaction is shown in Figure 17.

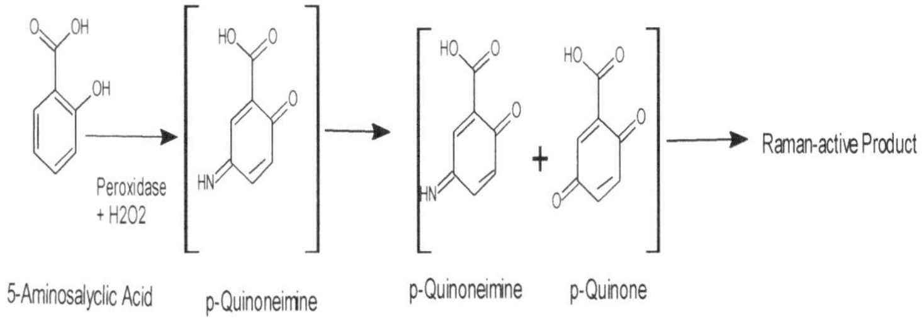

Recent Advances in Immunoassays

5-Aminosalyclic Acid   p-Quinoneimine   p-Quinoneimine   p-Quinone   Raman-active Product

Figure 17. Principle of Sword Raman Reaction with Peroxidase Substrate

The performance of this detection methodology using free peroxidase is compared to commonly used colorimetric, fluorescent and chemilumiescent detection methodologies. The mean response values were plotted against the HRPO standard concentrations on semi-log plots. The data was then fit to a 4PLC. These fitted curves were graphed on these plots to demonstrate the relative fit of the observed data as the plots provided in Figure 18 to demonstrate. The signals from all detection methodologies (Sword substrate, TMB as the colorimetric substrate, two separate fluorescence substrates and a chemiluminescence substrate) were drawn of the sample plot, scaling the response axis (Y axis) such that the Negative

Recent Advances in Immunoassays

Control (0 pg/mL) and highest positive (3,000 pg/mL) values from each respective curve overlapped. The data obtained using all three detection reagents yield smooth dose response curves that fit well to a 4PLC curve (with R^2 values of more than 0.9996 for all 4PLC fits (131).

More importantly, the Sword Peroxidase Reagent derived dose response curves are shifted to the left, relative to the colorimetric, fluorescent and chemiluminescent reagent derived curves. The magnitude of this shift is described by evaluating the concentration at ½ the maximum response of the highest calibrator (Conc.$^{1/2Max}$). This suggests that the Sword Peroxidase Reagents are capable of detecting lower concentrations (concentrations often of interest in assay) and improved sample distinguishably from the Negative Control samples. The shifted curves yielded an improvement in the estimated Analytical Limit of Detection (LOD) observed (Table 5). This demonstrates that Sword Detection System is about 2-100 fold more sensitive than these commercially available

Recent Advances in Immunoassays

substrates (129, 130, 132). The increase in leftward shift for dose response curves with Sword detection system compared with the most sensitive colorimetric, fluorescents and chemiluminescent substrates is also demonstrated in Figure 18.

Table 5. Analytical Sensitivity of Peroxidase with Different Detection Methodologies

Detection Methodology	Vendor	LOD (pg/mL)	Conc. $^{1/2Max}$
Sword Raman	Sword Diagnostics	2.7	510
Colorimetric (TMB)	KPL, Inc	16.8	1,534
Fluorescent (Amplex® Red)	Invitrogen	4.4	1,171
Fluorescent (QuantaRed™)	Thermo Fisher	284	1,928
Chemiluminescence (LumiGLO®)	KPL, Inc	15.8	2,437
Chemiluminescence (SuperSignal®)	Thermo Fisher	6.21	2,631

Recent Advances in Immunoassays

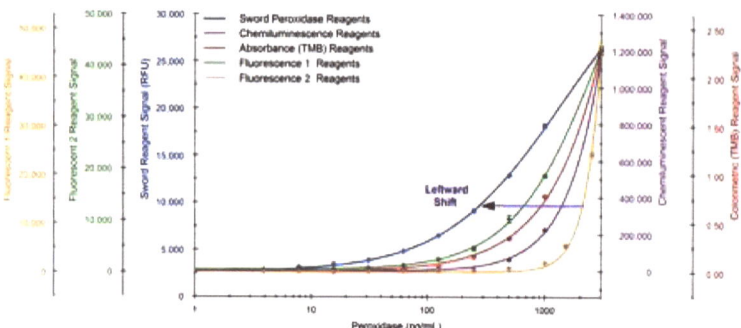

Figure18. Peroxidase Dose Response Curves with Colorimetric (TMB), Fluorescent 1 (Amplex® Red), Fluorescent 2 (QuantaRed™) and Chemiluminescent (LumiGLO®)Substrates

The Sword Detection Reagent was compared with the very sensitive chemiluminescent peroxidase substrate SuperSignal® (Thermo Fisher) and the most sensitive colorimetric TMB (KPL, Inc) with peroxidase. The dose response curves of these substrates are shown in Figure 19.

Recent Advances in Immunoassays

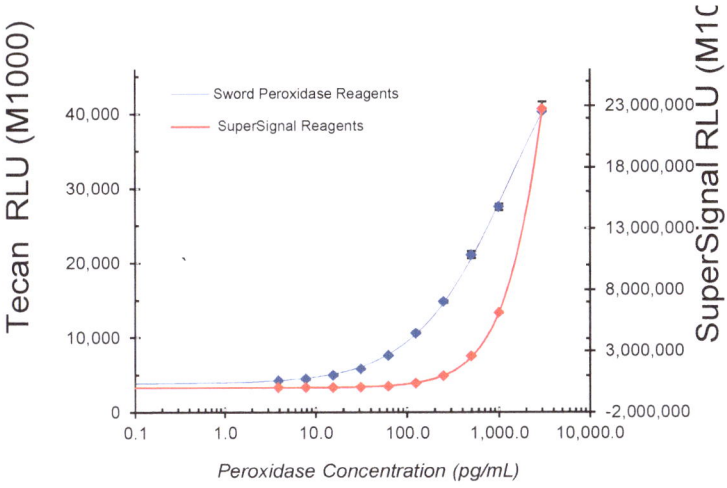

Figure 19, Dose Response Curves of Sword and SuperSignal® Reagents

The sensitivity of the assay which is the Analytical Limit of Detection (LOD) and concentrations at ½ Max from the assay specific dose response curves fitted with the 4PLC equation are shown in Table 6. These results demonstrate that the dose response curves for the Sword reagents were shifted leftwards (toward lower analyte concentrations) compared to native substrates. This shift is indicative of an increased detection capability of low analyte samples with Sword detection reagents relative to that observed with

101

Recent Advances in Immunoassays

the native assays tested in this study. Dose response curves of Human TNF-~and IL-6 assays are shown in Figure 20 and Figure 21 to illustrate these curve shifts (133-135). The sensitivity enhancements were further demonstrated with Human C-reactive protein ELISA and a Mouse IL-6 ELISA (136, 137).

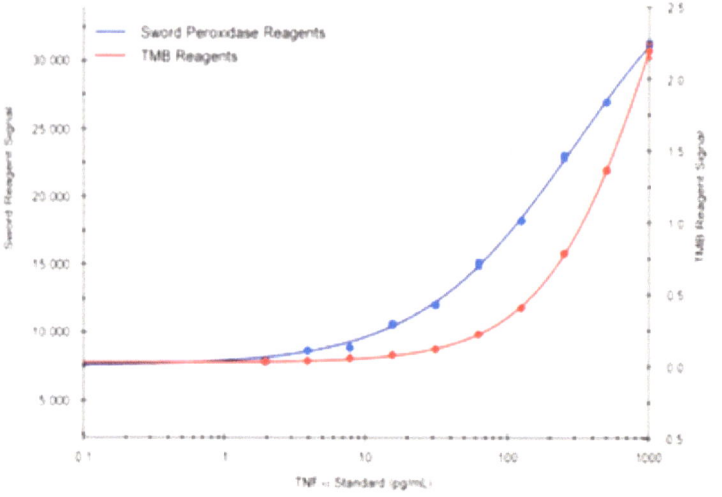

Figure 20. Dose Response Curves of Human TNF-α with Sword and TMB Reagents

Recent Advances in Immunoassays

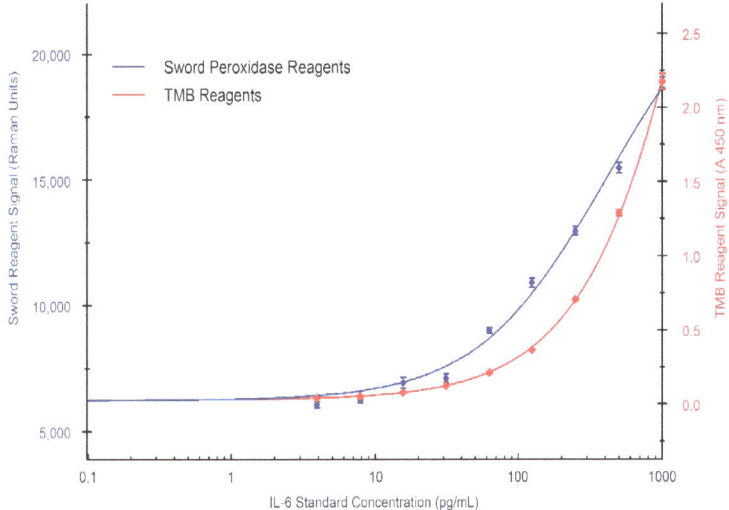

Figure 21. Dose Response Curves of Human IL-6 with Sword and TMB Reagents

Recent Advances in Immunoassays

Table 6   Analytical Sensitivity of Assays as Measured with Sword Peroxidase Reagents

Assay	Limit of Detection (pg/mL)				Curve Shift (Conc $^{1/2\ Max}$) In pg/mL
	Package Insert	TMB	Sword	Improvement to TMB	Improvement Relative to TMB
HRPO	NA	9.4	1.1	8.6 fold	3 fold
Human IL-2	7	5	0.4	12.7 fold	3 fold
Human IL-6	0.70	1.3	0.038	34.2 fold	ND
Mouse IL-6	NA	1.76	0.37	4.8 fold	4.1 fold
Human IL-8	5	3.6	1.3	2.8 fold	9 fold
Human VEGF	8	9.3	3.5	2.6 fold	94 fold
Human TNF-α	1.7	2.35	0.44	2.6 fold	5.2 fold
Human sTNF RII	0.60	0.51	0.24	2.0 fold	5.9 fold
Human IFN-γ	2	7	4.7	1.5 fold	2 fold
Human CRP	10	21.37	6.21	3.4 fold	5.6 fold

The enhanced sensitivity of the Sword detection reagent compared to the most sensitive colorimetric substrate TMB was illustrated further with a Human TNF-α ELISA kit by reduction of sample volumes. This is shown in Figure 22. The dose response curve for TMB remains at the extreme right and even with 4-fold reduction of sample, Sword reagent still remains towards the left, indicating more than 4-fold increased sensitivity.

104

Recent Advances in Immunoassays

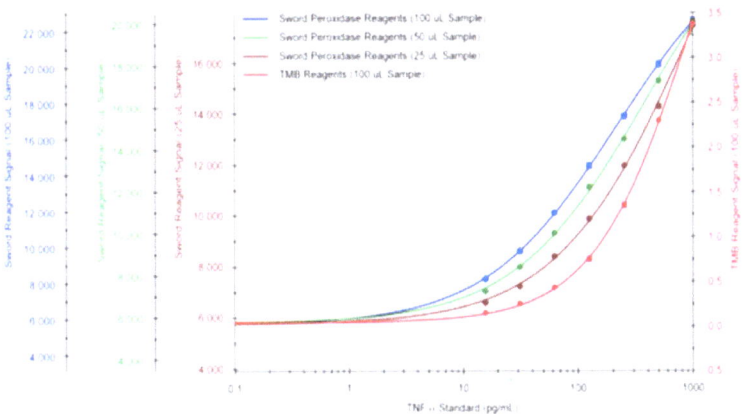

Figure 21. Dose Response Curves with Human TNF α. Note TMB curve remains at the right even the sample volume is reduced 4-fold

Use of Raman based detection in peroxidase-based immunoassays has been shown to enhance the sensitivity of existing assays using colorimetric, fluorescent and chemiluminescent peroxidase substrates available commercially. This detection system can be easily incorporated into existing assays without modification to the assay format, or the use of a dedicated microplate reader. The detection reagents are quite robust and no special environmental protection is required (as opposed to many commercially offered detection reagents). The leftward shift in

Recent Advances in Immunoassays

the dose response curves with these reagents often results in an immediate sensitivity improvement, beneficial for the detection of low abundance biomarkers. Follow-up optimization often results in performance enhancement, as demonstrated with the mouse IL-6 assays. Similar follow-up optimization improvements have been observed using commercially available human TNF-~kits and the Sword detection reagents (data not shown).

Application of Sword Raman Detection With Alkaline Phosphatase

Sword Detection reagents have been modified to allow the detection and quantitation of another marker enzyme Alkaline Phosphatase (ALP) used in variety of immunoassays. In the presence of this enzyme, phosphorylated aromatic phenol or aminophenol-derived substrates are enzymatically dephosphorylated to form the corresponding phenol or aminophenol. These compounds are easily oxidized to generate the corresponding Raman-active quinones or iminoquinones. The

Recent Advances in Immunoassays

appearance of these Raman active compounds can be analytically correlated to the amount Alkaline Phosphatase present. On this basis, the phosphorylated derivates of these phenolic compounds can be very useful as Alkaline Phosphatase substrates in immunoassays. Two examples are given here to emphasize the potential of developing very high sensitivity immunoassays with Alkaline Phosphatase using these highly detectable phenolic compounds.

Two compounds were examined. The first, 4-Aminophenyl phosphate (AAP) (Alexis Biochemicals) is known to undergo catalytic dephosphorylation by Alkaline phosphatase to yield 4-aminophenol. This compound has been used Two compounds were examined. The first, 4-Aminophenyl phosphate (AAP) (Alexis Biochemicals) is known to undergo catalytic dephosphorylation by Alkaline phosphatase to yield 4-aminophenol. This compound has been used in highly sensitive electrochemical immunoassays (138). The second compound, 4-

Recent Advances in Immunoassays

Aminonaphthyl phosphate (ANP) is known to undergo catalytic dephosphorylation by Alkaline Phosphatase to yield 4-amino-1-naphthol. The 4-Aminonaphthyl phosphate was synthesized as described by M. Masson et. al. for use in this evaluation. This compound has been used in amperometric immunoassays with high sensitivity (96) These compounds both rapidly oxidize to 1, 4-iminoquinone and 1, 4-iminonaphthaquinone respectively, under alkaline conditions (97).

To evaluate use of these compounds in Raman based assays, Calf Intestine Alkaline Phosphatase (Sigma Chemicals) or goat anti-human IgG Alkaline Phosphatase Conjugate (KPL Laboratories) were added as test samples to substrate dissolved in an alkaline buffer at pH 9.8. The reactions were incubated at room temperature for 30 or 60 minutes for the AAP and ANP substrates respectively. The Raman spectra were evaluated at the end of each reaction. Unlike the Horseradish Peroxidase substrate, sodium

Recent Advances in Immunoassays

hydroxide was not added to obtain the Raman signal.

The Raman spectra obtained using these substrates, like those observed with the Horseradish Peroxidase substrate, were broad in nature, and were the same whether free or conjugated Alkaline Phosphatase participated in the catalytic reaction. Not surprisingly these spectra are distinct from each other and from that obtained with the 5-Aminosalicyclic acid. It was noted though that the signal obtained using the 4-Aminonaphthyl phosphate was roughly an order of magnitude greater than that seen with the 4-Aminophenyl phosphate substrate.

In order to quantify levels of bacterial Alkaline Phosphatase (Sigma-Aldrich Chemicals), Raman signal was measured at 1,842 cm^{-1}. The Raman assay was compared with either an Alkaline Phosphatase colorimetric assay using p-Nitrophenylphosphate (pNPP) (Sigma-Aldich Chemicals), BluePhos (KPL, Inc) or an Alkaline Phosphatase fluorescent assay using 4-

Recent Advances in Immunoassays

Methylumbelliferyl phosphate (MUP) (Sigma Aldrich Chemicals).

Dose response curves, correlating the concentration of Alkaline Phosphatase with the measured Raman signal were obtained. These curves were similar to the dose response curves observed with Horseradish Peroxidase substrate (data not shown). The Raman responses are compared against pNPP and MUP responses in this evaluation. The leftward shift of the ANP Raman dose response curve (toward lower analyte concentrations) compared to either the pNPP and BluePhos colorimetric assay or the MUP fluorescence assay is indicative of an increased detection capability of low analyte samples with Raman detection reagents. This is further elucidated when the Sample / Negative (S/N) ratio values are examined. There was a substantial increase in the S/N ratio values (approximately 200 - 300 % increase with Alkaline Phosphatase concentrations greater than 1 mU/mL) with the ANP substrate compared with the

Recent Advances in Immunoassays

pNPP, BluePhos colorimetric assay or the MUP fluorescence assay (data not shown). The purposes of this evaluation was to demonstrate the feasibility of applying modified detection reagents to allow the Raman quantitation of a second commonly used detection enzyme. However, the use of Raman Alkaline Phosphate detection needs to be fully demonstrated with 4-Aminonaphthylphosphate (ANP) and 4-Hydroxynahthylphosphate (HNP) which are not yet commercially available (139).

Enzyme immunoassays of different categories have been used but they are not widely used. Some of them are Capillary Electrophoresis Immunoassay (4, 140, 141), Liposome Immunoassay (4, 24, 25, 142) Flow Injection Immunoassay (4,143,144) and Biosensors (145-148).Although enzymes have been outstandingly useful for signal amplification, they have disadvantages, including their molecular size and complexity, possible instability, and dependence of catalytic rate on temperature and other variables.

17.   Acknowledgments

I gratefully acknowledge Sword Diagnostics for helping me writing the book. My acknowledgement is to Kevin Oliver of Perkin Elmer for permission to include the diagram in Figure 6 and Figure 11; to Joe Barco of Singulex for permission to include the diagram in Figure 7; to Tim Huang of LSI Medience Corporation (Mitsubishi Group) for permission to include the diagram in Figure 10; to Becton and Dickenson for courtesy to include the diagram in Figure 14; to Christine Vale of Luminex for the permission to include the diagram Figure 15, and to Tina Cuccia of Bio-Rad for the permission to include the diagram in Figure 16.

18.   References

1. The Immunoassay Handbook, Editor: David Wild, Elsevier (2005)
2. Immunoassay, Editors: E P Diamandis and T. K. Christopolus, Elsevier (2013)
3. Andreotti, P E, Ludwig, G V, Anne Harwood Peruski, A H, J. Tuite, J J, Morse S S and Leonard F. Peruski, L F Jr, Immunoassay of Infectious agents, BioTechniques 35:850-859 (2003)
4. Ibrahim D, A Immunoassay Methods and their Applications in Pharmaceutical Analysis: Basic Methodology and Recent Advances, International Journal of Biomedical science, 217-235 (2006)
5. Tetin, S Y and Stroupe, S D, Antibodies In Diagnostic applications Current Pharmaceutical Biotechnology, 5; 9-16 (2004)
6. Weeks I, Chemiluminescence Immunoassay In Editor: G. Sevehla , Elsevier, 24; 8-118 (1992).
7. Simmonet F, Guilloteau D In editors R. F. Masseyeff, Albert W H, Staines N A Method of Immunological Analysis.New York: VCH.1; 270-282 (1993)
8. Porstsmann B, Portsmann T, Non-isotopic Immunoassay In Editor: T. T. Ngo. New York: Plenum Press, 60 (1988)
9. Lanthanide Luminescence: Photophysical, Analytical and Biological Aspects. P. Hannienan and H. Harana, Springer-Verlag (2011)
10. Soini E and Kojola H , Time-resolved fluorometer for lanthanide chelates-a new of non- isotopic immunoassays. Clinical

Recent Advances in Immunoassays

Chemistry, 29; 65-68 (1983)

11. Hemmila I, Fluoroimmunoassays an immuno-fluorometric assays. Clinical Chemistry 31; 359- 370 (1985)

12. Siitari H, Hemmila I, Lovgren T, Koistinnen V, Detection of hepatitis B surface antigen using time-resolved fluoroimmunoassay. Nature 301: 258-280 (1983)

13. Kricka, L J, Chemiluminescent and Bioluminescent technques. Clinical Chemistry 37:1472-1481(1991)

14. Whitehead, T P, Kricka, L J, Carter, J N, Thorpe, G, Analytical Luminescence: Its Potential in the Clinical Laboratory, Clinical Chemistry, 25: (1979)

15. Sakamaki N , Ohiro Y, Ito M, Makinodan M , Ohta O, Suzuki W, Takayasu S and Tsuge, Bioluminescent Enzyme Immunoassay for the Detection of Norovirus Capsid Antigen Clinical Vaccine Immunol. 19; 1949 (2012)

16. Weeks I, Chemiluminescence Immunoassay In Wilson and Wilsons Comprehensive Analytical Chemistry, Editor: G. Svehla, Elsevier, 29 (1992)

17. Whitehead, T. P., Phenols as Enhancers of the Chemiluminescent Horseradish Peroxidase-Luminol-hydrogen Peroxide Reaction: Application In Luminescence-Monitored Enzyme Immunoassays, Clinical Chemistry 31; 1335-1341 (1985).

18. Thorpe, G H G., Kricka, L J, Enhanced Chemluminescent Reactions Catalyzed by Horseradish Peroxidase, Methods in Enzymology, 133: 331-353 (1986)

Recent Advances in Immunoassays

19. Bronstein, I, Edwards, B, Voyta, J C, 1, 2-Dioxetanes: Novel Chemiluminescent Enzyme Substrates. Applications to Immunoassays, Journal of Bioluminescence and Chemiluminescence, 4; 99-111 (1989).

20. Beck, S, Koster, H, Applications of Dioxetane Chemiluminescent Probes to Molecular Biology, Analytical Chemistry, 62; 2258-2270 (1990).

21. Tizard, R, Cate, R L, Ramachandran, K L, Wysk, M, Voyta, J C, Murphy, O J, Bronstein, I, Imaging of DNA Sequences with Chemiluminescence, Proc. Natl. Acad. Sci. USA, 87, 4514-4518 (1990)

22. W, Nchimi N K, Syrjänpää M, Charbonnière LJ, Hovinen J, Härmä H. Stable and highly fluorescent europium(III) chelates for time- resolved immunoassays. Inorg Chem. , 52; 8461- 8466 (2013)ang Q

23. Nishioka T, Yuan J, Yamamoto Y, Sumitomo Wang Z, Hashino K, Hosoya C, Ikiawa K, Wang G, New luminescent europium(III) chelates for DNA labeling. Inorg Chem.,45 :4088-4096 (2006)

24. Gerber M A, Randolph, M F and DeMeo K K Liposome immunoassay for rapid identification of group A streptococci directly from throat swabs Clin Microbiol.,28; f 1463–1464 (1990)

25. Ahn-Yoon S, DeCory T R, Baeumner A J, Durst R A, Ganglioside-liposome immunoassay for the ultrasensitive detection of cholera toxin. Anal. Chem. 75 , 2256-2261 (2003)

26. Thermo Fisher (Pierce) Immunoassay Handbook

Recent Advances in Immunoassays

27. Bidwell, D and Bartlett, A. Enzyme- linked Immunosorbent assay, p. 359-371 In Editors: N.R. Rose and H. Friedman, Manual of Clinical Immunology. American Society of Microbiology, Washington, DC (1980)

28. Andreotti PE, Meyer R, Campbell T, Goode M T, Menking D L, Myers E D, Palenius T and Stanker L. Detection and identification of animal and food pathogens using time-resolved fluorescence, p. 857 In Editors :Y.-R. Chen and S.-I. Tu Photonic Detection and Intervention Technologies for Safe Food Proceedings of SPIE (2001)

29. TuS –I, Golden M C and Andreotti P E, Simultaneous detection of *E. coli* O157:H7 and Salmonella typhimurium in foods using time-resolved fluorescence of lanthanides In Proceedings of United States-Japan Natural Resources Panel. Sukuba, Japan (2001)

30. Engvall E. and Perlmann P. Enzyme-linked immunosorbent assay (ELISA). Quantitative assay of immunoglobulin G. Immunochemistry 8:871-874 (1971)

31. Engvall E. and Perlmann P. Enzyme-linked immunosorbent assay, ELISA Quantitation of specific antibodies by enzyme-labeled anti-immunoglobulin in antigen-coated tubes. J. Immunol. 109:129-135 (1972)

32. Plaksin D and Gromakovski E. White Paper, PolyHRP Detection: a real Technology for Ultrasensitive immunoassay Stereospecific Detection Technologies (SDT, Germany).

33. Charbgoo F, Mirshahi M, Sarikhani S, Saifi A M Synthesis of a unique high-performance poly-horseradish peroxidase complex to enhance

sensitivity of immunodetection systems, Biotechnol Appl Biochem, 59; 45-49 (2012)

34. Checovich W J, Bolger R E and Burke T. Fluorescence polarization- a new tool for cell and molecular biology, Nature 375; 254- 256 (1995)

35. Tencza SB, Islam K R, Kalia V, Nasir M S, Jolley M E and Montelaro, R C, Development of a fluorescence polarization-based diagnostic assay for equine infectious anemia virus, J. Clinical Microbiol., 38; 1854-1859 (2000)

36. LeTilly V and Royer C A, Fluorescence anisotropy assays implicate protein-protein interactions in regulating trp repressor DNA binding. Biochemistry 32; 7753-7758 (1993)

37. Nielsen K, Gall, D Jolley M, Leishman M, Balsevicius G, Smith G, Nicoletti P and Thomas F, A homogeneous fluorescence polarization assay for detection of antibody to Brucella abortus. J. Immunol. Methods 195; 161-168 (1996)

38. Maeda H. Assay of proteolytic enzymes by the fluorescence polarization technique. Anal. Biochem. 92; 222-227 (1979)

39. Bolger R F, Lenoch E, Allen B ,Meiklejohn B and Burke T, Fluorescent dye assay for detection of DNA In recombinant protein products. BioTechniques 23; 532-537 (1997).

40. Singh, K, Rucker, T Hanne,A, Parwaresch, R and Krupp, G, Fluorescence polarization for monitoring ribozyme reactions in real time. BioTechniques 29; 344-351 (2000)

41, Hagan K and Zuchner, T. Lanthanide-based time-resolved luminescence immunoassays, Anal Bioanal Chem. 400; 2847–2864 (2011)

42. Dodeigne C, Thunus C L, Lejeune R. Chemilumi-nescence as diagnostic tool. A review, Talanta 51; 415–439 (2000)

43. Applied Biosystems (Foster City, CA) Product Guide Chemiluminescent Substrates and Enhancers), LD Publication 120CA01-01 (2007)

44. Weeks I and Woodhead J S. Chemiluminescent assays based on acridinium labels, p. 553-556 In Editors: P.E. Stanley and L.J. Kricka, Biolumi-nescence and Chemiluminescence: Current Status. John Wiley & Sons, Chichester, England (1991)

45. Albrecht S, Ehle H, Schollberg K, Bublitz R, Horn A. Chemiluminescent enzyme immunoassay of human growth hormone based on adamantly dioxetane phenyl phosphate substrate, Bioluminescence and Chemiluminescence: Current Status, p 115-118. (1991)

46. Bronstein I, Edwards B, Voyta J C. 1,2-dioxetanes: novel chemiluminescent enzyme substrates: Applications to immunoassays, J. Biolum. Chemilum. 4; 99-111(1989)

47. Bronstein I, Voyta J C, Thorpe GH, Kricka L J, Armstrong G. Chemiluminescent assay of alkaline phosphatase applied in an ultrasensitive enzyme immunoassay of thyrotropin, Clinical Chemistry 35; 1441-1446 (1989)

48. Bronstein I, Olesen C E M, Martin C, Schneider G, Edwards B, Sparks A, and Voyta J C (1994) Chemiluminescent detection of DNA and protein with CDP and CDP-Star 1,2-dioxetane enzyme substrates, Bioluminescence and Chemilumi-

nescence: Fundamentals and Applied Aspects. pp. 269-272 (1994)

49. Beck, S., Koster, H., Applications of Dioxetane Chemiluminescent Probes to Molecular Biology, Analytical Chemistry 62; 2258-2270 (1990).

50. Bioluminescence and Chemiluminescence: Fundamentals and Applied Aspects. Campbell A K, Kricka L J and Stanley P E, editors. John Wiley & Sons, Chichester, England. p. 321-324 (1994)

51. Invitrogen (Carlsbad, CA), Chemiluminescent Alkaline Phosphatase ELISA Systems, Technical Bulletin MP 10552, June 2009, References cited in.

52. Szekeres P, Leong K, Day T, Kingston A, and Karran E. Development of homogeneous 384-well high-throughput screening assaysfor Ab 1-40 and Ab 1-42 using AlphaScreen technology. J Biomol Screen. 13; 101-111 (2008)

53. Sehr P, Pawlita M, and Lewis J. Evaluation of different glutathione S-transferase-tagged protein captures for screening E6/E6AP interaction nhibitors using AlphaScreen. J Biomol Screen.12; 560-567(2007)

54. Poulsen F and Jensen K B. A luminescent oxygen channeling immunoassay for the determination of insulin in human plasma. J Biomol Screen.12; 240- 247 (2007)

55. Wilson J, Rossi C P, Carboni S, Fremaux C, Perrin D, Soto C, Kosco M and Scheer A. A homogeneous 384-well high-throughput binding assay fora TNF receptor using AlphaScreen technology. J Biomol Screen.8; 522-532 (2003)

Recent Advances in Immunoassays

56. Shukla R, Santoro J, Bender F, Latzera. Quantitative determination of human interleukin 22 (IL-22) in serum using Singulex-Erenna® Technology, Journal of Immunological Methods, 390: 30–34 (2013)

57. Wu A H B, Agee S J, Anh Q A, Todd J, Specicificity of a High-Sensitivity Cardiac Troponin I Assay Using Single-Molecule-counting Technology. Clinical Chemistry, 55; 196-198 (2009)

58. Ledger K S, Agee S J ,Kasaian M T, Forlow S B Durn B L, Minyard J, Lu Q A, Todd J, Vesterqvist O , Burczynski M E, Analytical validation of a highly sensitive microparticle-based immunoassay for the quantitation of IL-13 in human serum using the Erenna® immunoassay system. J Imunol Methods. 350; 161-170 (2009)

59. Ho J E, Hwang S J, Wollert K C, Larson M G, Kempf T, Vasan R S, Januzzi J L, Wang T J, and Fox C S. Biomarkers of Cardiovascular Stress and Incident Chronic Kidney Disease, Clinical Chemistry Published July 19, 2013, 205716

60. Knight A W. A review of recent trends in analytical applications of electro-generated chemiluminescence. Trends Anal. Chem., 18; 47- 62 (1999)

61. WIKI Books, Analytical Chemiluminescence/ Electrochemiluminescence

Recent Advances in Immunoassays

62. Higgins J A, Ibrahim M S, Knauert F K, Ludwig G V, Kijek T M , Ezzell J W, Courtney B C , and Henchal E A. Sensitive and rapid identification of biological threat agents. Ann. NY Acad. Sci. 894;130-148 (1999)

63. Kijek, TM, Rossi C A, Moss D, Parker R W, and Henchal E A. Rapid and sensitive immunomagnetic-electrochemiluminescent detection of Staphyloccocal enterotoxin B. J. Immunol. Methods. 236; 9-17 (2000)

64. Yang H, Leland J K D, Yost D and R J. Massey R J. Electrochemiluminescence: a new diagnostic and research tool. ECL detection technology promises scientists new "yardsticks" for quantification. Biotechnology (NY) 12; 193-194 (1994)

65. Gowan S M, et al. Application of Meso Scale Technology for the measurements of phosphor-proteins -in human xenografts, Assay and Drug Dev Technol, 5; 391-401 (2007)

66. Thompson L, et al. Competitive electrochemi-luminescence (ECL) was and no-wash immunoassays for the detection of serum antibodies to smooth Brucella strains, Clin Vaccine Immunol. 16; 765-771 (2009)

67. Cludts, I, et al. Detection of neutralizing interleukin-17 antibodies in autoimmune

Recent Advances in Immunoassays

polyendocrinopathy syndrome-1 (APS-1) patients using a novel non-cell based electrochemiluminescence assay. Cytokine 50;129-137 (2010)

68. Lembo A, et al. Use of serum biomarkers in a diagnostic test for irritable bowel syndrome. Aliment Pharmacol Ther. 29; 834-842 (2009)

69. Leng S X, McElhaney J E, Walston J D, Xie D,Fedarko N S and Kuchel G A. ELISA AND MULTIPLEX TECHNOLOGIES FOR CYTOKINE MEASUREMENT IN INFLAMMATION AND AGING RESEARCH J Gerontol A Biol Sci Med Sci. 63; 879–884 (2008)

70. Moxness, M, et al., Immunogenicity Testing by Electrochemiluminescent Detection for antibodies directed against therapeutic human monoclonal antibodies. Clinical Chemistry, 51;1983-1985 (2005)

71. Lateral Flow Immunoassay, editors Raphael C. Wong, I Harley Y. Tse, Humana Press Springer (2009)

72. Arai H, Petchclai B , Khupulsup K, Kurimura T, and Takeda K. Evaluation of a Rapid Immuno-chromatographic Test for Detection of Antibodie to Human Immunodeficiency Virus, J Clin Microbiol. 37; 367–370 (1999)

73. Herring A J, Ballard R C, Pope V, Adegbola R A, Changalucha J, Fitzgerald D W, Hook E W III, Kubanova A, Mananwatte S, Pape J W, Sturm A W, West B, Yin Y P, Peeling R W. A multi-centre evaluation of nine rapid, point-of-care Syphilis tests using archived sera. SexTransm Infect. 82; v7–v12 (2006)

Recent Advances in Immunoassays

74. Zuk R F, Ginsberg V K, Gouts T, Rabbie J, Merrick H, Ullman E F, Fischer M M , Sizto C C et al, Enzyme immunochromatography - a quantitative imunoassay requiring no instrumentation. Clinical Chemistry 31; 1144-1150 (1985)
75. Corstjens, P, Zuiderwijk M, Brink A, Li S, Feindt H, Niedbala R S , and Tanke H. Use of up-converting phosphor reporters in lateral-flow assays to detect specific nucleic acid sequences: a rapid, sensitive DNA test to identify human papillomavirus type 16 infection, Clinical Chemistry 47; 1885-1893 (2001)
76. Niedbala, R S, Feindt H, Kardos K, Vail T, Burton   J, Bielska B, Li S, Milunic D, et al. Detection of analytes by immunoassay using up-converting phosphor technology. Anal.Biochem.293: 22-30 (2001)
77. Papadea C, Foster J, Grant S, Ballard S A, Cate J C IV, W. Southgate M, and Dilip M. Purohit D M. Evaluation of the i-STAT Portable Clinical Analyzer for Point-of-Care Blood Testing in the Intensive Care Units of a University Children's Hospital, Annals of Clinical & Laboratory Science, 32; 231-243 (2002)
78. Erickson K A, Wilding P. Evaluation of a novel point of care system, the i-STAT portable clinical analyzer. Clinical Chemistry, 39; 283-287 (1993)
79. Murthy J N, Hicks J M, Soldin S J. Evaluation of i-STAT portable clinical analyzer in a neonatal and pediatric intensive care unit. Clin Biochem, 30; 385- 389 (1997)
80. Bingham D, Kendall J, Clancy M. The portable laboratory: and evaluation of the accuracy and reproducibility of i-STAT. Ann Clin Biochem,36;

66- 71 (1999).

81. Jacobs E, Vadasdi E, Sarkozi L, Colman N. Analytical evaluation of i-STAT portable clinical analyzer and use by non-laboratory health-care professionals. Clinical Chemistry, 39; 1069-

82. Palamalai V, Murakami M M, Apple F S. Diagnostic Performance of four point of care cardiac troponin I assays to rule in and rule out acute myocardial infarction, Clin. Biochem.,46; 1631-5 (2013)

83. Hierholzer, JC, Johansson K H, Anderson L J, Tsou C J, and Halonen P E. Comparison of monoclonal time-resolved fluoroimmunoassay with monoclonal capture-biotinylated detector enzyme immunoassay for adenovirus antigen detection. Clin. Microbiol. 25; 1662-1667 (1987)

84 Hemmila, I S, Dakubu V M, Mukkala H, Siitari, Lovgren T. Europium as a label in time-resolved immunofluorometric assays. Anal.Biochem., 137; 335-343 (1984)

85 Fernandez S M, Wang H P, Chao Y S, Guignon E F. TIME-RESOLVED FLUORESCENCE United States Patent Number: 4,923,819 (1993)

86. Smith, D.R. et al. Comparison of Dissociation-Enhanced Lanthanide Fluorescent imunoassays to Enzyme-Linked Immunosorbent Assays for Detection of Staphylococcal Enterotoxin B, Yersinia pestis-Specific F1 Antigen, and Venezuelan Equine Encephalitis Virus. Clin Diagn Lab Immun., 8; 1070-1075 (2001)

87. Hu, Z. et al. Detection of hepatitis B virus PreS1 antigen using a time-resolved fluoro-immunoassay. J Immunoassay Immunochem., Immunochem., 33; 156–165 (2012).

Recent Advances in Immunoassays

88. Hemmilä I, Holttinen S, Pettersson K and Lovgren T. Double-Label Time-Resolved Immunofluorometry of Lutropin and Follitropin in Serum. Clinical Chemistry 33; 2281-2283 (1987).

89. Xu, Y.Y. et al. Simultaneous Quadruple-Label Fluorometric Immunoassay of Thyroid-Stimulating Hormone, 17 Alpha Hydroxy-progesterone, Immunoreactive Trypsin, and Creatine Kinase MM Isoenzyme in Dried Blood Spots. Clinical Chemistry 38; 2038-2043 (1992)

90. Gyros: Miniaturizing immunoassays for improved performance in Application Report 207).

91. Lakkis M, Kai J, Santiago N , Puntambekar A, Moore V, Lee S H, Sehy D W, Schultheis R, J Han J and Ahn C H. Novel Biomarkers Detection and Identification by Microfluidic-Based MicroELISA, Translational Medic 2012, S:1 pp 1-5 (2012)

92. Kai J, Santiago N, Puntambekar A, Lee S H, Sehy D, Schultheis R, Han J, Ahn C H, Brescia P and Banks P. Amplifying Immunoassay Sensitivity with the Optimiser™ Microplate Technology, BioTek Instruments, Inc., Winooski, VT Application Notes, (2011)

93. Kai J, Moore V, Puntambekar A, Sehy D, Lakkis M, Lee S, Han J, and Ahn C. Optimiser™ Microplate enables simultaneous multi- analyte quantitation using ultra-low sample volumes in a 2-hour assay: Validation of a Novel 10-Analyte Th17 Cell Panel. The Journal of Immunology, 188; 124-127 (2012)

Recent Advances in Immunoassays

94. Diaz-Gonzalez, Gonzalez-Garcia M B and Costa- Garcia A. "Recent advances in electrochemichal enzyme immunoassays," Electroanalysis, 17; 1901–1918 (2005).

95. Taleat Z, Cristea C, Marrazza G, and Sndulescu R. Electrochemical Sandwich Immunoassay for the Ultrasensitive Detection of Human MUC1 Cancer Biomarker.International Journal of Electrochemistry, Article ID 740265, pp. 1-6 (2013).

96. Masson M, Runaround O, Johannson F and Airzawa M. 4-Amino-1-naphthylphosphate as a substrate for the amperometric detection of alkaline phosphatase activity and its application for immunoassay. Talenta 64; 74-180 (2004)

97. Masson M, Haruyama T, Koobatake E and Airzawa 4-Hydroxynaphthyl-1-phosphate as a substrate for alkaline phosphatase and its use in sandwich immunoassay. Analytica Chimica Acta 402: 29-35 (2004).

98. Cook E B, Stahl J L, Lowe L, Chen R, Morgan E, Wilson J, Varro R, Chan A, Graziano F M, Barney N P. Simultaneous measurement of six cytokines in a single sample of human tears using microparticle-based flow cytometry: allergics vs. nonallergics. J Immunol Methods 254:109–18 (2001)

99. Yan X, Schielke E G, Grace K M, Hassell C, Marrone B L, Nolan J P. Microsphere-based duplexed immunoassay for influenza virus typing by flow cytometry. J Immunol Methods 284; 27–38 (2004)

100. Aoe K, Hiraki A, Murakami T, Murakami K, Makihata K, Takao K, Eda R, Maeda T, Sugi K, Darzynkiewicz Z, Takeyama H. Relative abundance and patterns of correlation among six

Recent Advances in Immunoassays

cytokines in pleural fluid measured by cytometric bead array. Int J Mol Med $\underline{12}$:193–8 (2003)

101. Kellar KL, Douglass JP. Multiplexed Microsphere-based Flow Cytometric Immunoassays for Human Cytokines. J Immunol Meth.,50:239–242 (2003).

102. Chen R, Lowe L, Wilson J D, Crowther E, Tzeggai K, Bishop J E, Varro R. Simultaneous Quantification of Six Human Cytokines in a Single Sample Using Microparticle-based Flow Cytometric Technology. Clinical Chemistry $\underline{45}$; 1693–1694 (1999)

103. Khan S S, Smith M S, Reda D, Suffredini A F, McCoy J P Jr. Multiplex bead array assays for detection of soluble cytokines: comparisons of sensitivity and quantitative values among kits from ultiple manufacturers. Cytometry B Clin Cytom $\underline{64}$; 53 (2005).

104. Young S H, Antonini J M, Roberts J R, Erdley A D, Zeidler-Erdeley P C. Performance evaluation of cytometric bead assays for the measurement of lung cytokines in two rodent models. J of Immunological Methods, $\underline{331;}$ 59–68 (2008)

105. Earley M C, Vogt R F, Shapiro H M, Mandy F F, Kellar K L, Bellisario R, Pass K A, Marti G E, Stewart C C, Hannon W H. Report from a Workshop on Multianalyte Microsphere Arrays.Cytometry50; 239–242 (2002)

106. Kettman J R, Davies T, Chandler D, Oliver K G, Fulton R J. Classification and properties of 64 multiplexed microsphere sets. Cytometry $\underline{33}$; 234– 243 (1998)

107. de Jager W, te Velthuis H, Prakken BJ, Kuis W, Rijkers GT. Simultaneous detection of 15 human cytokines in a single sample of stimulated peripheral blood mononuclear cells. Clin Diagn

Recent Advances in Immunoassays

Lab Immunol 10; 133–139 (2003)

108. Liu M Y, Xydakis A M, Hoogeveen R C, Jones P H, Smith E O, Nelson K W, Ballantyne C M. Multiplexed analysis of biomarkers related to obesity and the metabolic syndrome in human plasma, using the Luminex-100 system. Clinical Chemistry. 51; 1102–1109 (2005).

109. Keyes K, Mann L, Cox K, Treadway P, Iverson P, Chen Y F, Teicher B A. Circulating angiogenic growth factor levels in mice bearing human tumors using Luminex Multiplex technology. Cancer Chemother Pharmacol,51; 321-327 (2003)

110. Brett Houser. Bio-Rad's Bio-Plex® suspension array system, xMAP technology Overview Arch Physiol Biochem. 118: 192–196 (2012)

111. Matthew J. Binnicker, Deborah J. Jespersen, and Leonard O. Rollins, Evaluation of the Bio-Rad BioPlex Measles, Mumps, Rubella, and Varicella-Zoster Virus IgG Multiplex. Bead Immunoassay Clin Vaccine Immunol.18: 1524–1526 (2011)

112. Michael J. Benecky, Diane R. Post, Susan M. Schmitt, and Manish S. Kochar, Detection of hepatitis B surface antigen in whole blood by coupled particle light scattering (Copalis™). Clinical Chemistry 43; 1764–1770 (1997)

113. NICOLA BIZZARO, FABRIZIO BONELLI, ELIO TONUTTI, RENATO TOZZOLI, AND DANILO VILLALT. A New Coupled-Particle Light-Scattering Assay for Detection of Ro/SSA (52 and 60 Kilodaltons) and La/SSB Autoantibodies in Connective Tissue Diseases. .CLINICAL AND DIAGNOSTIC LABORATORY IMMUNOLOGY pp. 922–925 (2001)

Recent Advances in Immunoassays

114. Nulens E, Bodeus M, Bonelli F, Soleti A, Goubau P. Reactivity to p52 and CM2 recombinant proteins in primary human cytomegalovirus infection with a microparticle agglutination assay. Clin DiagnLab Immunol 7; 536–9 (2000)

115. Larsen, S A.; Hunter, E F.; Kraus, S J., A manual of tests for syphilis. American Public Health Association; Washington, DC. pp. 191(1990)

116. Pang S, Smith J, Onley D, Reeve J, Walker M, Foy C. A Comparability Study of the Emerging Protein Array Platforms with Established ELISA Procedures. J Immunol Methods 302;1–12 (2005).

117. Prabhakar U, Eirikis E, Davis HM. Simultaneous Quantitation of Proinflammatory Cytokines in Human Plasma Using the abMAP Assay. J Immunol Methods 260; 207–218 (2002)

118. Hildesheim A, Ryan R L, Rinehart E, Nayak S, Wallace D, Castle P E, Niwa S, Kopp W. Simultaneous measurement of several cytokines using small volumes of biospecimens. Cancer Epidemiol Biomarkers Prev 11:1477–84 (2002)

119. DuPont N C, Wang K, Wadhwa P D, Culhane J F, Nelson EL. Validation and Comparison of Luminex Multiplex Cytokine Analysis Kits with ELISA: Determinations of a Panel of Nine Cytokines in Clinical sample Culture Supernatants. J Reprod Immunol 66; 175–191 (2005)

120. Ray C A, Bowsher R R, Smith W C, Devanarayan V, Willey M B, Brandt J T, Dean R A. Development, Validation, and Implementation of a Multiplex Immunoassay for the Simultaneous Determination of Five Cytokines

in Human Serum. J Pharm Biomed Analysis. 36; 1037–1044 (2005)

121. Elshal M F and McCoy J P Jr. Multiplex Bead Array Assays: Performance Evaluation and Comparison of Sensitivity to ELISA Methods 38; 317–323 (2006).

122. Carson RT and Vignali D A. Simultaneous quantitation of 15 cytokines using a multiplexed flow cytometric assay. J. Immunol. Methods 227; 41-52 (1999).

123. de Koning L, Liptakc C, Shkreata A, Bradwin G, Huf F B , PradhanA D, Rifai N,Kellog M D. A multiplex immunoassay gives different results than Singleplex immunoassays which may bias epidemiologic associations, Clinical Biochem. 45: 848 (2012)

124. Breen E C, Reynolds S M, Cox C, Jacobson L P, Magpantay L, Mulder C B, Dibben O, Margolick J B, Bream J H, Sambrano E, Martínez-Maza O, Sinclair E, Borrow P, Landay A L, Rinaldo C R, and Norris P J. Multisite Comparison of High-Sensitivity Multiplex Cytokine Assays. CLINICAL AND VACCINE IMMUNOLOGY, pp. 1229–1242 (2011)

125. Oliver K G, Kettman J R, Fulton R J. Multiplexed Analysis of Human Cytokines by Use of the FlowMetrix System. Clinical Chemistry 44; 2057– 2060 (1998).

126. Elshal MF, McCoy JP. Multiplex bead array assays: performance evaluation and comparison of sensitivity to ELISA. Methods 38; 317–323 (2006)

127. Scaife P J, Innes B A, Otun H A, Robson S C, Searle R F, Bulmer J N. Comparison of three multiplex cytokine analysis systems: Luminex,

Recent Advances in Immunoassays

SearchLight™ and FAST Quant® J of
Immunological Methods <u>309</u>; 205–208 (2006)

128. Nyquist N et al, "Handbook of Infrared and
Raman Spectra of Inorganic Compounds and
Organic Salts", Vol. 1-4 (New York: Academic
Press, 1996)

129. Kundu S, Siegel N, Ginsburgh C, Martin O,
Chokshi K and Lynch S. Raman detection
system for immunoassays: A sensitive Raman-
scattering method for peroxidase based
enzyme-linked immunoassays. IVD Technology
Magazine, Canon Communication Publications,
summer, (2010)

130. Kundu S, Ginsburgh C, Lynch S, and Siegel N.
Ultrasensitive Immunoassay Detection System
for Biomarkers Utilizing Raman-   Scattering
Methods. American Laboratory Technical
Publication, November (2012)

131. Rodbard D. Statistical quality control and
routine data processing for radioimmunoassays
and immunoradiometric assays. Clinical
Chemistry <u>20</u>; 1255–1270 (1974)

132. Peroxidase Detection using Sword™
Diagnostics Peroxidase Reagents: Enhanced
Performance using the Tecan Infinite ™ M1000
multimode reader. Tecan Technical Note 396612
V 1.0, (Tecan Group Ltd, Sword Diagnostics)
(2011)

133. Human TNF-α ELISA using Sword™
Diagnostics Peroxidase Reagents: Enhanced

Recent Advances in Immunoassays

Performance using the Tecan Infinite ™ M1000 multimode reader. Tecan Technical Note 396610 V 1.0, (Tecan Group Ltd, Sword Diagnostics) (2011)

134. High Sensitivity Human IL-6 ELISA with Expanded Dynamic Range ELISA using Sword™ Diagnostics Peroxidase Reagents: Enhanced Performance using the Tecan Infinite™ M1000 multimode reader. Sword Technical Note (www.sworddiagnostics.com)

135. Human IL-6 chemiluminescent ELISA using Sword™ Diagnostics Peroxidase Reagents: Enhanced Performance using the Tecan Infinite™ M1000 multimode reader. Tecan Technical Note 396608 V 1.0, 02-2011 (Tecan Group Ltd, Sword Diagnostics)

136. Human C-reactive protein ELISA using Sword™ Diagnostics Peroxidase Reagents: Enhanced:Performance using the Tecan Infinite™ M1000 multimode reader. Tecan Technical Note 396606 V 1.0, 02-2011 (Tecan Ltd, Sword Diagnostics)

137. An Optimized Mouse IL-6 ELISA using B using Sword™ Diagnostics Peroxidase Reagents: Enhanced Performance using the Tecan Infinite

Recent Advances in Immunoassays

™ M1000 multimode reader. Tecan Technical
Note 396728 V 1.0, 10-02-2011 (Tecan Group
Ltd, Sword Diagnostics)

138. Tang, H. T. et al. "p-Aminophenylphosphate: an
improved substrate for Electrochemical enzyme
immunoassay" Analyica Chimica Acta 214:
90-95(1988)

139. Siegel N S, Kundu SK and Ginsburgh C L.
METHODS FOR DETECTING ORGANISMS
AND ENZYMATIC REACTIONS USING
RAMAN SPECTROSCOPY AND AROMATIC
COMPOUNDS COMPRISING PHOSPHATE:
US Patent US 2010/0041016 A1 (2010)

140. Wang Q, Guoan Luo L, Wang Y and Yeung W S
B. Capillary Electrophoresis Base
Immunoassay for Monoclonal Antibody with
Diode Laser Induced Fluorescence Detection.
Analytical Letters 33, pp. 589-602 (2000)

141. Moser A C, Hage D S. Capillary
electrophoresis- based immunoassays:
principles and quantitative applications.
Electrophoresis 29; 3279-95 (2008

142. Kim C-K and Lim S-J. Liposome immunoassay
(LIA) with antigen-coupled liposomes containing
alkaline phosphatase. J Immunological
Methods. 59: 101–106 (1993)

143. Hansen E H Principles and applications of flow
injection analysis in biosensors. J Mol Recognit:

Recent Advances in Immunoassays

$\underline{9}$: 316-325 (1996)

144. Baeyens W R, Schukman S G, Calokerinos A C, Zhao Y, Garcia Campana A M, Nakashima K, de Keukeleire D. Chemiluminescence-based detection: principles and analytical applications in flowing streams and in immunoassays. J Pharm Biomed Anal.17: 941-53 (1998)

145. García-Campaña A M, Baeyens W R, Zhang X R, Smet E, Van Der Weken G, Nakashima K, Calokerinos A C. Detection in the liquid phase applying chemiluminescence . Biomed. Chromatogr., $\underline{14}$: 166-72 (2000)

146. Shan G, Lipton C, Gee S J and Hammock B D. Immunoassay, biosensors and other Non-chromatographic methods In Handbook of Residue Analytical Methods for Agrochemicals Editor Philip W. Lee, John Wiley & Sons, Ltd, Chichester, England, pp. 623–679 (2002)

147. Zhang J, Zhu J, Xu J, Chen H, and Xu D, "An Electrochemical impedimetric arrayed immunosensor based on indium tin oxide electrodes and silver-enhanced gold nanoparticles, Microchimica Acta, $\underline{163}$; 63–70 (2008).

148. Shao Y, Wang J, Wu H, Liu J, Aksay I A, Lina Y. Graphene Based Electrochemical Sensors and Biosensors: A Review Electroanalysis $\underline{22}$: 1027 – 1036 (2010)

www.ingramcontent.com/pod-product-compliance
Lightning Source LLC
Chambersburg PA
CBHW040806200526
45159CB00022B/31